Annals of Mathematics Studies

Number 54

ELEMENTARY DIFFERENTIAL TOPOLOGY

BY

James R. Munkres

Lectures

Given at Massachusetts Institute of Technology

Fall, 1961

REVISED EDITION

PRINCETON, NEW JERSEY

PRINCETON UNIVERSITY PRESS

1966

Printed in the United States of America
Revised edn., 1966
Second Printing, with corrections, 1968
Third Printing, 1973

To the memory of
Sumner B. Myers

PREFACE

Differential topology may be defined as the study of those properties of differentiable manifolds which are invariant under differentiable homeomorphisms. Problems in this field arise from the interplay between the topological, combinatorial, and differentiable structures of a manifold. They do not, however, involve such notions as connections, geodesics, curvature, and the like; in this way the subject may be distinguished from differential geometry.

One particular flowering of the subject took place in the 1930's, with work of H. Whitney, S. S. Cairns, and J. H. C. Whitehead. A second flowering has come more recently, with the exciting work of J. Milnor, R. Thom, S. Smale, M. Kervaire, and others. The later work depends on the earlier, of course, but differs from it in many ways, most particularly in the extent to which it uses the results and methods of algebraic topology. The earlier work is more exclusively geometric in nature, and is thus in some sense more elementary.

One may make an analogy with the discipline of Number Theory, in which a theorem is called elementary if its proof involves no use of the theory of functions of a complex variable — otherwise the proof is said to be non-elementary. As one is well aware, the terminology does not reflect the difficulty of the proof in question, the elementary proofs often being harder than the others.

It is in a similar sense that we speak of the elementary part of differential topology. This is the subject of the present set of notes.

Since our theorems and proofs (with one small exception) will involve no algebraic topology, the background we expect of the reader consists of a working knowledge of: the calculus of functions of several variables and the associated linear algebra, point-set topology, and, for Chapter II, the geometry (not the algebra) of simplicial complexes. Apart from these topics, the present notes endeavor to be self-contained.

The reader will not find them especially elegant, however. We are not hoping to write anything like the definitive work, even on the most elementary aspects of the subject. Rather our hope is to provide a set of notes from which the student may acquire a feeling for differential topology, at least in its geometric aspects. For this purpose, it is necessary that the student work diligently through the exercises and problems scattered throughout the notes; they were chosen with this object in mind.

The word problem is used to label an exercise for which either the result itself, or the proof, is of particular interest or difficulty. Even the best student will find some challenges in the set of problems. Those problems and exercises which are not essential to the logical continuity of the subject are marked with an asterisk.

A second object of these notes is to provide, in more accessible form than heretofore, proofs of a few of the basic often-used-but-seldom-proved facts about differentiable manifolds. Treated in the first chapter are the body of theorems which state, roughly speaking, that any result which holds for manifolds and maps which are infinitely differentiable holds also if lesser degrees of differentiability are assumed. Proofs of these theorems have been part of the "folk-literature" for some time; only recently has anyone written them down. ([8] and [9].) (The stronger theorems of Whitney, concerning analytic manifolds, require quite different proofs, which appear in his classical paper [15].)

In a sense these results are negative, for they declare that nothing really interesting occurs between manifolds of class C^1 and those of class C^∞. However, they are still worth proving, at least partly for the techniques involved.

The second chapter is devoted to proving the existence and uniqueness of a smooth triangulation of a differentiable manifold. In this, we follow J. H. C. Whitehead [14], with some modifications. The result itself is one of the most useful tools of differential topology, while the techniques involved are essential to anyone studying both combinatorial and differentiable structures on a manifold. The reader whose primary interest is in triangulations may omit §4, §5, and §6 with little loss of continuity.

We have made a conscious effort to avoid any more overlap with the lectures on differential topology [4] given by Milnor at Princeton in 1958 than was necessary. It is for this reason that we omit a proof of Whitney's imbedding theorem, contenting ourselves with a weaker one. We hope the reader will find our notes and Milnor's to be useful supplements to each other.

Remarks on the revised edition

Besides correcting a number of errors of the first edition, we have also simplified a few of the proofs.

In addition, in Section 2 we now prove rather than merely quote the requisite theorem from dimension theory, since few students have it as part of their background. The reference to Hurewicz and Wallman's book was inadequate anyway, since they dealt only with finite coverings.

Finally, we have added at the end some additional problems which exploit the Whitehead triangulation techniques. The results they state have already proved useful in the study of combinatorial and differentiable structures on manifolds.

CONTENTS

ELEMENTARY DIFFERENTIAL TOPOLOGY

CHAPTER I.

DIFFERENTIABLE MANIFOLDS

§1. Introduction.

This section is devoted to defining such basic concepts as those of differentiable manifold, differentiable map, immersion, imbedding, and diffeomorphism, and to proving the implicit function theorem.

We consider the euclidean space R^m as the space of all infinite sequences of real numbers, $x = (x^1, x^2, \ldots)$, such that $x^i = 0$ for $i > m$; euclidean half-space H^m is the subset of R^m for which $x^m \geq 0$. Then $R^{m-1} \subset H^m \subset R^m$. We denote $\sqrt{((x^1)^2 + \ldots + (x^m)^2)}$ by $\|x\|$, and $\max |x^i|$, by $|x|$. The unit sphere S^{m-1} is the subset of R^m with $\|x\| = 1$; the unit ball B^m, the set with $\|x\| \leq 1$; and the r-cube $C^m(r)$ is the set with $|x| \leq r$. Often, we also consider R^m as simply the space of all m-tuples $(x^1, \ldots, x^m)$, where no confusion will arise.

1.1 Definition. A (topological) manifold M is a Hausdorff space with a countable basis, satisfying the following condition: There is an integer m such that each point of M has a neighborhood homeomorphic with an open subset of H^m or of R^m.

If $h : U \to H^m$ (or R^m) is a homeomorphism of the neighborhood U of x with an open set in H^m or R^m, the pair (U,h) is often called a coordinate neighborhood on M. If h(U) is open in H^m and h(x) lies in R^{m-1}, then x is called a boundary point of M, and the set of all such points is called the boundary of M, denoted by Bd M. If Bd M is empty, we say M is non-bounded. (In the literature, the word manifold is commonly used only when Bd M is empty; the more inclusive term then is manifold-with-boundary.) The set $M - \text{Bd}\, M$ is called the interior of M, and is denoted by $\text{Int}\, M$. (If A is a subset of the topological space X, we also use $\text{Int}\, A$ to mean $X - C\ell(X-A)$, but this should cause no confusion.)

To justify these definitions, we must note that if $h_1 : U_1 \to H^m$ and $h_2 : U_2 \to H^m$ are homeomorphisms of neighborhoods of x with open sets in H^m, and if $h_1(x)$ lies in R^{m-1}, so does $h_2(x)$: For otherwise, the map $h_1 h_2^{-1}$ would give a homeomorphism of an open set in R^m with a neighborhood of the point $p = h_1(x)$ in H^m. The latter neighborhood is certainly not open in R^m, contradicting the Brouwer theorem on invariance of domain [3, p. 95].

One may also verify that the number m is uniquely determined by M; it is called the dimension of M, and M is called an m-manifold. This may be done either by using the Brouwer theorem on invariance of domain, or by applying the theorem of dimension theory which states that the topological dimension of M is m [3, p. 46]. Strictly speaking, to apply the latter theorem we need to know that M is a separable metrizable space; but this follows from a standard metrization theorem of point-set topology [2, p. 75].

Because M is locally compact, separable, and metric, M is paracompact [2, p. 79]. We remind the reader that this means that for any open covering $\mathfrak{A}$ of M, there is another such collection $\mathfrak{B}$ of open sets covering M such that

(1) The collection $\mathfrak{B}$ is a refinement of the first, i.e., every element of $\mathfrak{B}$ is contained in an element of $\mathfrak{A}$.

(2) The collection $\mathfrak{B}$ is locally-finite, i.e., every point of M has a neighborhood intersecting only finitely many elements of $\mathfrak{B}$.

In passing, let us note that because M has a countable basis, any locally-finite open covering of M must be countable.

(a) Exercise. If M is an m-manifold, show that $\text{Bd}\, M$ is a non-bounded $m-1$ manifold or is empty.

(b) Exercise. Let M be an m-manifold with non-empty boundary. Let $M_0 = M \times 0$ and $M_1 = M \times 1$ be two copies of M. The double of M, denoted by $D(M)$, is the topological space obtained from $M_0 \cup M_1$ by identifying $(x,0)$ with $(x,1)$ for each x in Bd M. Prove that $D(M)$ is a non-bounded manifold of dimension m.

(c) Exercise. If M and N are manifolds of dimensions m and n, respectively, then $M \times N$ is a manifold of dimension $m + n$, and $\mathrm{Bd}(M \times N) = ((\mathrm{Bd}\ M) \times N) \cup (M \times (\mathrm{Bd}\ N))$.

1.2 Definition. If U is an open subset of R^m, then $f : U \to R^n$ is differentiable of class C^r if the partial derivatives of the component functions $f^1, \ldots, f^n$ through order r are continuous on U. If f is of class C^r for all finite r, it is said to be of class C^∞.

If A is any subset of R^m, then $f : A \to R^n$ is differentiable of class C^r $(1 \leq r \leq \infty)$ if for each point x of A, there is a neighborhood U_x of x such that $f|(A \cap U_x)$ may be extended to a function which is of class C^r on U_x.

If $f : A \to R^n$ is differentiable, and x is in A, we use $Df(x)$ to denote the Jacobian matrix of f at x — the matrix whose general entry is $a_{ij} = \partial f^i/\partial x^j$. We also use the notation $\partial f^1, \ldots, f^n/\partial(x^1, \ldots, x^m)$ for this matrix. Now f must be extended to a neighborhood of x before these partials are defined; in the cases of interest, the partials are independent of the choice of extension (see Exercise (b)).

We recall here the chain rule for derivatives, which states that $D(fg) = Df \cdot Dg$, where fg is the composite function, and the dot indicates matrix multiplication.

(a) Exercise. Check that differentiability is well-defined; i.e., that the differentiability of $f : A \to R^n$ does not depend on which "containing space" R^m for A is chosen.

(b) Exercise. Let U be open in R^m; let $U \subset A \subset \bar{U}$. If $f : A \to R^n$ is of class C^1, and x is in A, show that $Df(x)$ is independent of the extension of f to a neighborhood of x in R^m which is chosen.

Remark. Our definition of the differentiability of a map $f: A \to R^n$ is essentially a local one. We now obtain an equivalent global formulation of differentiability, which is given in Theorem 1.5.

1.3 Lemma. There is a C^∞ function $\varphi : R^m \to R^1$ which equals 1 on $C(1/2)$, is positive on the interior of $C(1)$, and is zero outside $C(1)$.

Proof. Let $f(t) = e^{-1/t}$ for $t > 0$, and $f(t) = 0$ for $t \leq 0$. Then f is a C^∞ function which is positive for $t > 0$.

Let $g(t) = f(t)/(f(t) + f(1-t))$. Then g is a C^∞ function such that $g(t) = 0$ for $t \leq 0$, $g'(t) > 0$ for $0 < t < 1$, and $g(t) = 1$ for $t \geq 1$.

Let $h(t) = g(2t+2)\, g(-2t + 2)$. Then h is a C^∞ function such that $h(t) = 0$ for $|t| \geq 1$, $h(t) > 0$ for $|t| < 1$, and $h(t) = 1$ for $|t| \leq 1/2$.

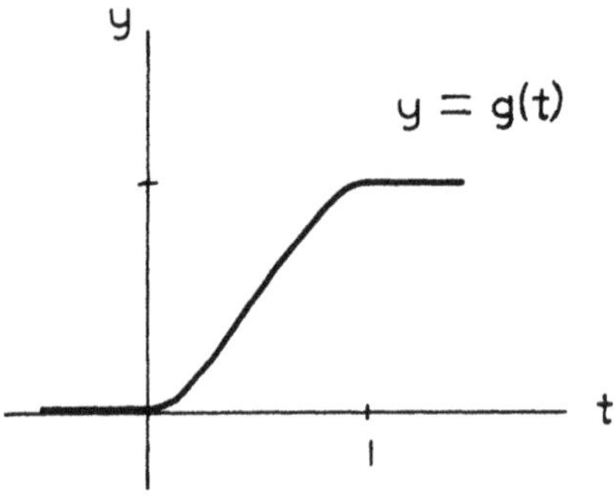

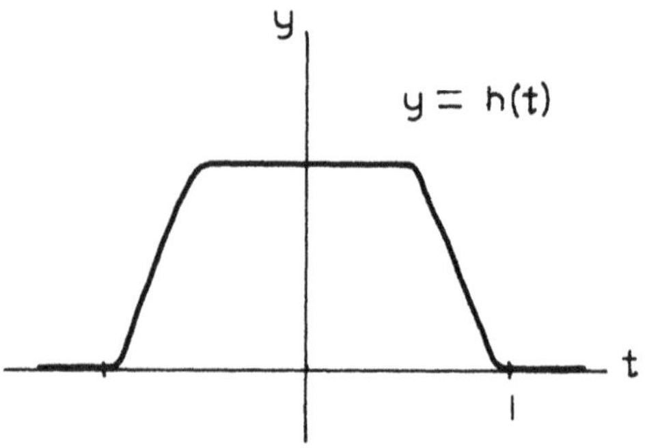

Let $\varphi(x^1,\ldots,x^m) = h(x^1) \cdot h(x^2) \cdots h(x^m)$.

(a) Exercise. Generalize the preceding lemma as follows: Let U be an open subset of R^m; let C be a compact subset of U. There is a C^r real-valued function Ψ defined on R^m such that Ψ is positive on C and is zero in a neighborhood of the complement of U.

Remark. Whenever an indexed collection $\{C_i\}$ of subsets of X is said to be locally-finite, we shall mean by this that every point of X has a neighborhood intersecting C_i for at most finitely many values of i. This convention is convenient; it implies that only the empty set can equal C_i for more than finitely many i.

1.4 Lemma. (1) Let $\{U_i\}$ be an open covering of the paracompact space X. There is a locally-finite open covering $\{V_i\}$ of X such that $V_i \subset U_i$ for each i. (2) Let $\{V_i\}$ be a locally-finite open covering of the normal space X, where $i = 1, 2, \ldots$, there is a closed covering (C_i) of X such that $C_i \subset V_i$ for each i.

Proof. (1) Let $\mathcal{B}$ be a locally-finite refinement of $\{U_i\}$. For each element B of $\mathcal{B}$, choose an index $j(B)$ such that $B \subset U_{j(B)}$. Define V_j as the union of those B for which $j(B) = j$. Each point has a neighborhood intersecting only finitely many elements of $\mathcal{B}$; this neighborhood intersects V_j for only finitely many values of j.

(2) We construct the coverings $\{C_i\}$ by induction. Let W_1 be an open set containing $X - (V_2 \cup V_3 \cup \ldots)$, whose closure is contained in V_1. Let $C_1 = \bar{W}_1$.

Suppose $W_1 \cup \ldots \cup W_{j-1} \cup V_j \cup V_{j+1} \cup \ldots = X$. Let W_j be an open set containing

$$X - (W_1 \cup \ldots \cup W_{j-1} \cup V_{j+1} \cup \ldots)$$

whose closure is contained in V_j. Let $C_j = \overline{W}_j$.

To prove that the collection $\{W_j\}$ covers X, note that any point x lies in only finitely many sets V_j. Hence for some j, x is not in $V_j \cup V_{j+1} \cup \ldots$. As a result, x must belong to $W_1 \cup \ldots \cup W_{j-1}$, by the induction hypothesis. Transfinite induction may be used to prove (2) for an arbitrary index set.

1.5 Theorem. Let A be a subset of R^m; let $f : A \to R^n$ be of class C^r. Then f may be extended to a function which is of class C^r in a neighborhood of A.

Proof. By hypothesis, for each point x of A, there is a neighborhood U_x of x such that $f|A \cap U_x$ may be extended to a function which is of class C^r on U_x. We choose $\bar{U}_x$ to be compact. Let M be the union of the sets U_x; it is an open subset of R^m. Let $\{V_i\}$ be a locally-finite open refinement of the covering $\{U_x\}$ of M. Let $\{C_i\}$ be a covering of M by closed sets such that $C_i \subset V_i$ for each i. Let Ψ_i be a C^∞ function defined on R^m which is positive on C_i and equals zero in a neighborhood of the complement of V_i. Here we use Exercise (a) of 1.3. Then $\Sigma\, \Psi_j(x)$ is a C^∞ function on M, since it equals a finite sum in some neighborhood of any given point of M. Define $\varphi_i(x) = \Psi_i(x)/\Sigma\, \Psi_j(x)$; then $\Sigma\, \varphi_i(x) = 1$.

For each i, let f_i denote a C^r extension of $f|A \cap V_i$ to V_i; if $A \cap V_i$ is empty, let f_i be the zero function. Then $\varphi_i f_i$ may be extended to be of class C^r on M by letting it equal zero outside V_i. Define

$$\tilde{f}(x) = \Sigma_i\, \varphi_i(x)\, f_i(x) \quad .$$

This is a finite sum in some neighborhood of any point x of M, and hence is of class C^r on M. Furthermore, if x is in A, then $\varphi_i f_i = \varphi_i f$ for every i, so that

$$\tilde{f}(x) = \Sigma\, \varphi_i(x)\, f(x) = f(x) \;.$$

Hence $\tilde{f}$ is the required C^r extension of f to the neighborhood M of A in R^m.

1.6 Definition. A differentiable m-manifold of class C^r is an m-manifold M and a differentiable structure $\mathfrak{D}$ of class C^r on M. A differentiable structure of class C^r on M, in turn, is a collection of coordinate neighborhoods (U,h) on M, satisfying three conditions:

(1) The coordinate neighborhoods in $\mathfrak{D}$ cover M.

(2) If (U_1,h_1) and (U_2,h_2) belong to $\mathfrak{D}$, then

$$h_1 h_2^{-1} \colon h_2(U_1 \cap U_2) \to R^m$$

is differentiable of class C^r.

(3) The collection $\mathfrak{D}$ is maximal with respect to property (2); i.e., if any coordinate neighborhood not in $\mathfrak{D}$ is adjoined to the collection $\mathfrak{D}$,

then property (2) fails.

The elements of $\mathfrak{D}$ are often called <u>coordinate systems</u> on the differentiable manifold M.

(a) Exercise. Let $\mathfrak{D}'$ be a collection of coordinate neighborhoods on M satisfying (1) and (2). Prove there is a unique differentiable structure $\mathfrak{D}$ of class C^r containing $\mathfrak{D}'$. (We call $\mathfrak{D}'$ a <u>basis</u> for $\mathfrak{D}$, by analogy with the relation between a basis for a topology and the topology.)

Hint: Let $\mathfrak{D}$ consist of all coordinate neighborhoods (U,h) on M which overlap every element of $\mathfrak{D}'$ differentiably with class C^r; this means that for each element (U_1,h_1) of $\mathfrak{D}'$,

$$h_1h^{-1} : h(U \cap U_1) \to R^m$$

and

$$h\,h_1^{-1} : h_1(U \cap U_1) \to R^m$$

are differentiable of class C^r. To show $\mathfrak{D}$ is a differentiable structure, the local formulation of differentiability is needed.

(b) Exercise. Let M be a differentiable manifold of class C^r (we often suppress mention of the differentiable structure $\mathfrak{D}$, where no confusion will arise). Then M may also be considered to be a differentiable manifold of class C^{r-1}, in a natural way; one merely takes $\mathfrak{D}$ as a basis for a differentiable structure $\mathfrak{D}_1$ of class C^{r-1} on M. Verify that the inclusion $\mathfrak{D} \subset \mathfrak{D}_1$ is proper. This proves that the class C^r of a differentiable manifold is uniquely determined.

We see in this way that the class of a differentiable manifold M may be lowered as far as one likes merely by adding new coordinate systems to the differentiable structure. The reverse is also true, but it will require much work to prove.

(c) Exercise*. If M is a differentiable manifold, what are the difficulties involved in putting a differentiable structure on D(M)? (D(M) was defined in Exercise (b) of 1.1.)

<u>1.7 Definition</u>. Let M and N be differentiable manifolds, of dimensions m and n, respectively, and of class at least C^r. Let A be a subset of M and let $f : A \to N$; then f is said to be <u>of class</u> C^r if

for every pair (U,h) and (V,k) of coordinate systems of M and N, respectively, for which $f(A \cap U) \subset V$, the composite

$$kfh^{-1} : h(A \cap U) \to R^n$$

is of class C^r. (Note that a map of class C^2 is also of class C^1, although a manifold of class C^2 is not one of class C^1 until the differentiable structure is changed.)

The rank of f at the point p of M is the rank of $D(kfh^{-1})$ at $h(p)$, where (U, h) and (V, k) are coordinate systems about p and $f(p)$, respectively. This number is well-defined, provided there is an open subset W of M such that $W \subset A \subset \overline{W}$, for if (U_1, h_1) and (V_1, k_1) were other such coordinate systems, we would have

$$D(k_1 f h_1^{-1}) = D(k_1 k^{-1}) \cdot D(kfh^{-1}) \cdot D(hh_1^{-1}) \quad .$$

The requirements for a differentiable structure assure that $k_1 k^{-1}$ and kk_1^{-1} are both differentiable, so that $D(k_1 k^{-1})$ is non-singular, having $D(kk_1^{-1})$ as its inverse. Similarly, $D(hh_1^{-1})$ is non-singular, so $D(k_1 f h_1^{-1})$ and $D(kfh^{-1})$ have the same rank.

(a) Exercise. The standard C^∞ differentiable structure on R^m is that having as basis the single coordinate system $i : R^m \to R^m$. Similarly for H^m. Check that the definitions of differentiability given in 1.2 and 1.7 agree.

1.8 Definition. Let $f : M \to N$ be differentiable of class C^r; let M and N have dimensions m and n, respectively. If rank $f = m$ at each point p of M, f is said to be an immersion. If f is a homeomorphism (into) and is an immersion, it is called an imbedding. If f is a homeomorphism of M onto N and is an immersion, it is called diffeomorphism; of course, $m = n$ in this case.

(a) Exercise. Note that $\text{Bd}\, H^m = R^{m-1}$ and the inclusion $R^{m-1} \to H^m$ is an imbedding. Generalize this as follows: If M is a differentiable manifold of class C^r, then there is a unique differentiable structure of class C^r on $\text{Bd}\, M$ such that the inclusion $\text{Bd}\, M \to M$ is a C^r imbedding.

(b) Exercise. Show that the composition of two immersions is an immersion.

(c) Exercise. Let M and N have class C^r; let M be non-bounded. Construct a C^r differentiable structure on $M \times N$ such that the natural inclusions of M and N into $M \times N$ are imbeddings. Why do we require M to be non-bounded?

(d) Exercise*. Construct a C^∞ differentiable structure on $H^1 \times H^1$ so that the inclusion $i : H^1 \times H^1 \to R^2$ is differentiable. What are the difficulties involved in putting a differentiable structure on $M \times N$ in general?

(e) Exercise*. Construct a C^∞ immersion of S^1 into R^2 which carries S^1 into a figure eight. Can such an immersion be extended to an immersion of B^2 into R^2?

(f) Exercise*. Prove that two differentiable structures of class C^r on the topological manifold M are the same if and only if the identity map from each differentiable manifold to the other is a C^r diffeomorphism.

(g) Exercise*. Construct two different differentiable structures of class C^∞ on the manifold R^1, such that the resulting differentiable manifolds are diffeomorphic.

1.9 Problem.* Prove that any two differentiable structures on R^1 give manifolds which are diffeomorphic.

(It was long a classical problem to determine whether two different differentiable structures on the same topological m-manifold always gave rise to diffeomorphic differentiable manifolds. Recently, it has turned out that the answer is yes if $m \leq 3$ [10, 13], and no if $m = 7$ [5]. The answer is also yes for R^n, if $n \neq 4$ [16]. Other results are known, but much work remains to be done.)

1.10 Problem. If $f : M \to N$ is a diffeomorphism of class C^r, prove that f^{-1} is a diffeomorphism of class C^r.

Hint: Prove the result in the following steps. Let g map the open subset U of R^m into R^n.

(1) $Dg(x)$ exists if there is a matrix A and a matrix function $R(x_1)$ such that

$$g(x_1) - g(x) = A \cdot (x_1 - x) + R(x_1)'$$

(here g, x_1, x, and R are written as column matrices), and $R(x_1)/\|x_1 - x\| \to 0$ as $x_1 \to x$. It follows that $A = Dg(x)$.

(2) If g is of class C^1 on U and A is a compact subset of U, then

$$g(x_1) - g(x) = Dg(x) \cdot (x_1 - x) + R(x_1, x),$$

where $R(x_1, x)/\|x_1 - x\| \to 0$ uniformly as $\|x_1 - x\| \to 0$, for x_1 and x in A. (For later use, this is stated in more generality than is actually needed here.)

(3) If g is of class C^1 on U and has rank m at x, then g satisfies a Lipschitz condition: There is a $\delta > 0$ and constants m and M such that

$$0 < m < \|g(x_1) - g(x)\| / \|x_1 - x\| < M$$

for $0 < \|x_1 - x\| < \delta$.

(4) Let g be a homeomorphism of the open subset U of R^m onto an open set in R^m; let g be of class C^1 on U and have rank m at x. Then $D(g^{-1})$ exists at $g(x)$ and equals the inverse of $Dg(x)$.

Remark. Previously, we appealed to the Brouwer theorem on invariance of domain to prove that the boundary and dimension of a manifold are well-defined. If one restricts oneself to considering manifolds for which there exist differentiable structures (and it has recently been shown by Kervaire [17] that this is a real restriction), then this appeal may be to some extent avoided by use of the following theorem, which resembles the invariance of domain theorem.

1.11 Theorem. Let f be a C^1 mapping of the open subset U of R^m into R^m which has rank m at the point x. Then f is a homeomorphism of some neighborhood of x onto a neighborhood in R^m of $f(x)$.

Proof. We may assume that x and $f(x)$ are at the origin, and that $Df(0) = I$, the identity matrix. (Why?) Choose r small enough that the cube $C(r)$ lies in U, and so that the maximum entry of $Df(x) - I$ is within $1/2m$ of 0 for x in $C(r)$. If we let $g(x) = f(x) - x$, then for x_1 and x_2 in $C(r)$, it follows from the mean-value theorem that

$$|g(x_1) - g(x_2)| \leq (1/2)|x_1 - x_2|$$

and

$$|f(x_1) - f(x_2)| \geq (1/2)|x_1 - x_2| \quad .$$

First, we prove that every point y of $\operatorname{Int} C(r/2)$ is the image under f of a point x of $\operatorname{Int} C(r)$. To do this, let $x_0 = 0$, $x_1 = y$, and in general,

$$x_{n+1} = y - g(x_n) \quad ,$$

providing x_n lies in the domain of g. Note that $|x_{n+1} - x_n| = $ $= |g(x_n) - g(x_{n-1})| \leq (1/2)|x_n - x_{n-1}|$, so that

$$|x_{n+1} - x_n| \leq (1/2)^n|x_1 - x_0| = (1/2)^n|y| \quad .$$

Summing this inequality, we see that

$$|x_{n+1} - x_m| < (1/2)^m \, 2|y|$$

for $m \leq n$. In particular, $|x_{n+1} - x_0| < 2|y| < r$. Hence x_{n+1} necessarily lies in the domain of g if $x_0, \ldots, x_n$ do, so that in fact x_n is defined for all n. It also follows that the sequence x_n is a Cauchy sequence. Hence it converges, say to x. Then $|x - x_0| \leq 2|y|$, so that x is in $\operatorname{Int} C(r)$. Also, $x = y - g(x)$, so that $f(x) = y$, as desired.

Second, we note that there is only one point x of $C(r)$ such that $f(x) = y$. For the contrary would contradict the fact that $|f(x_1) - f(x_2)| \geq (1/2)|x_1 - x_2|$ for x_1 and x_2 in $C(r)$.

Third, we note that the inverse map $f^{-1}\colon \operatorname{Int} C(r/2) \to C(r)$ is continuous, since $|y_1 - y_2| \geq (1/2)|f^{-1}(y_1) - f^{-1}(y_2)|$, for y_1 and y_2 in $\operatorname{Int} C(r/2)$.

Hence f is a homeomorphism of the open set $f^{-1}(\operatorname{Int} C(r/2)) \cap \operatorname{Int} C(r)$ onto the open set $\operatorname{Int} C(r/2)$, as desired.

1.12 <u>Corollary</u> (Inverse function theorem). Let f map the open subset U of R^m into R^m; let f be of class C^r and have rank m at x. Then f is a diffeomorphism of class C^r of a neighborhood of x onto a neighborhood in R^m of $f(x)$.

Proof. Since we know f is a homeomorphism on some such neighborhood, we need merely restrict the neighborhood to a smaller one on which the rank of f is m. This is possible because the set of those points x for which $\det Df(x) \neq 0$ is open.

1.13 Corollary (Implicit function theorem). Let $(x^1, \ldots, x^n, y^1, \ldots, y^p)$ denote the general point of $R^n \times R^p$. Let f be a C^r map of a neighborhood U of 0 in $R^n \times R^p$ into R^p; let $f(0) = 0$ and let $\partial f^1, \ldots, f^p/\partial(y^1, \ldots, y^p)$ be non-singular at 0. There is a unique function g mapping a neighborhood of 0 in R^n into R^p such that $g(0) = 0$, and $f(x, g(x)) = 0$ for x in this neighborhood; g is of class C^r.

Proof. Define $F : U \to R^{n+p}$ by the equation

$$F(x,y) = (x^1,\ldots,x^n, f^1(x,y),\ldots,f^p(x,y)) .$$

Since $DF(x)$ has the form

$$\begin{bmatrix} I & 0 \\ \partial f/\partial x & \partial f/\partial y \end{bmatrix}$$

it is non-singular at the origin. Hence F has a local inverse G near 0. If we are to have $f(x,g(x)) = 0$, then we must have

$$F(x,g(x)) = (x,f(x,g(x))) = (x,0) ,$$

so that

$$GF(x,g(x)) = (x,g(x)) = G(x,0) .$$

Hence we must define $g(x) = \pi G(x,0)$, where π is the projection of $R^n \times R^p$ onto R^p. We leave it to the reader to check that this definition of $g(x)$ will satisfy the requirements of the theorem.

1.14 Corollary. Let U be an open subset of R^m, let $f : U \to R^n$ be a map of class C^r such that f has rank m at the origin, and $f(0) = 0$. There is a C^r diffeomorphism g of a neighborhood of the origin in R^n onto another such that

$$gf(x^1,\ldots,x^m) = (x^1,\ldots,x^m,0,\ldots,0) .$$

Proof. We may assume the submatrix $\partial f^1,\ldots,f^m/\partial(x^1,\ldots,x^m)$ of

Df is non-singular at 0, since this condition may be obtained by following f by a suitable non-singular linear map. Define $F : U \times R^{n-m} \to R^n$ by the equation

$F : U \times R^{n-m} \to R^n$ by the equation

$$F(x^1,\ldots,x^n) = f(x^1,\ldots,x^m) + (0,\ldots,0,x^{m+1},\ldots,x^n) \quad .$$

Then F has rank n at 0, for DF has the block form

$$\left[\begin{array}{c|c} \partial f/\partial x & \begin{matrix} 0 \\ \\ I \end{matrix} \end{array}\right] \quad .$$

Hence F has a local inverse g. Now g is a C^r diffeomorphism of a neighborhood of the origin in R^n onto itself, and

$$gf(x^1, \ldots,x^m) = g(F(x^1,\ldots,x^m,0,\ldots,0)) = (x^1,\ldots,x^m,0,\ldots,0) \ .$$

Note. The preceding corollary also holds if U is an open subset of H^m. One merely extends f to an open subset of R^m and applies the result just proved.

1.15 Problem.* Let U be an open subset of R^m, let $f: U \to R^n$ be a map of class C^r such that f has rank n at the origin, and $f(0) = 0$. There is a C^r diffeomorphism h of a neighborhood of the origin in R^m onto another such that

$$fh(x^1,\ldots, x^m) = (x^1,\ldots, x^n).$$

1.16 Problem.* Let U be an open subset of R^m, let $f: U \to R^n$ be a map of class C^r such that f has rank k in a neighborhood of the origin, and $f(0) = 0$. There are C^r diffeomorphisms g and h of neighborhoods of the origin in R^n and R^m respectively onto others such that

$$gfh(x^1,\ldots, x^m) = (x^1,\ldots, x^k, 0, \ldots, 0).$$

§2. Submanifolds and Imbeddings.

In this section we define what is meant by a submanifold of a differentiable manifold, and prove that every differentiable manifold is a submanifold of some euclidean space.

2.1 Definition. Let N be an n-manifold, of class at least C^r; let N' denote the manifold of class C^r obtaining by possibly adding to the collection of coordinate systems. Let P be a subset of $\operatorname{Int} N$. P is called a submanifold of N of class C^r if there is a covering of P by coordinate systems (U_i, h_i) of N' such that $h_i(U_i \cap P)$ is an open subset of H^p or R^p, for some fixed p.

One shows at once that the coordinate neighborhoods $(U_i \cap P, h_i)$ form a basis for a differentiable structure of class C^r on P, and that relative to this structure on P, the inclusion $i : P \to N$ is a C^r imbedding.

It is appropriate to note here that ours is a different usage of the term submanifold than that employed by differential geometers. For them, a submanifold is the image of a one-to-one immersion $f : P \to N$ of one manifold in another, rather than the image of an imbedding.

Note that under our definition, the number r is not uniquely determined by P. The real line R^1, for instance, considered as a subset of the C^∞ manifold R^2, is a submanifold of R^2 of class C^r for every r between one and infinity.

(a) Exercise. Show that B^m and S^{m-1} are C^∞ submanifolds of R^m.

(b) Exercise*. Let R^2 be given the natural C^∞ structure. Given r with $1 \leq r < \infty$, construct a subset of R^2 which is a C^r submanifold of R^2 but not a C^{r+1} submanifold of R^2.

2.2 Theorem. If $f : M \to \text{Int } N$ is a C^r imbedding, then $f(M)$ is a submanifold of N of class C^r.

Proof. Let p be a point of M; let (U, h) be a coordinate system about p; and let (V, k) be a coordinate system about $f(p)$. Assume $h(p)$ and $kf(p)$ are at the origin (why is this justifiable?). Using Corollary 1.14, there is a diffeomorphism g of a neighborhood of $kf(p)$ in R^n onto an open set in R^n such that

$$gkfh^{-1}(x^1, \ldots, x^m) = (x^1, \ldots, x^m, 0, \ldots, 0) .$$

The coordinate system (V', gk) is the required coordinate system about $f(p)$; it carries a neighborhood of $f(p)$ in $f(M)$ onto an open set in H^m or R^m.

(a) Exercise*. Show that this theorem fails if f is only a C^r homeomorphism.

2.3 Problem*. Let $f : R^n \to R^1$ be of class C^r; let P be the set of points x for which $f(x) = 0$; suppose $\text{grad } f = \partial f/\partial(x^1, \ldots, x^n)$ is non-zero at each point x of P. Prove that P is a non-bounded submanifold of R^n of class C^r and dimension $n-1$.

State and prove a generalization to the case in which f maps R^n into R^m.

Remark. The preceding sections suggest that a very natural way in which differentiable manifolds may occur is as submanifolds of some euclidean space. At once the question arises as to whether every differentiable manifold so occurs. The answer to this question is yes, as we shall now prove. First we treat a special case.

2.4 Theorem. Let M be a compact m-manifold of class C^r. Then there is a C^r imbedding of M in some euclidean space.

Proof. For each p in $\text{Int } M$, choose a coordinate system (U,h) about p such that $h(U)$ contains $C(1)$, and $h(p)$ lies at the origin. (Why is this possible?) If p is in $\text{Bd } M$, we require $h(U)$ to contain

$C(1) \cap H^m$. Let V denote $h^{-1}(\text{Int } C(1))$, and let W denote $h^{-1}(\text{Int } C(1/2))$. The sets W cover M; choose a finite subcollection $W_1, \ldots, W_n$ which cover M; index the corresponding h, U, and V's accordingly.

Let φ be the function defined in Lemma 1.3. For $i = 1, \ldots, n$, define

$$\varphi_i(x) = \varphi(h_i(x)) \quad \text{for } x \text{ in } U;$$
$$\varphi_i(x) = 0 \quad \text{for } x \text{ in } M - \bar{V}.$$

Each map $\varphi_i : M \to R^1$ is of class C^∞, for it is of class C^∞ on the two open sets U and $M - \bar{V}$, and is well-defined on their intersection. Let $\Phi : M \to R^n$ be defined by the equation $\Phi(x) = (\varphi_1(x), \ldots, \varphi_n(x))$.

Define $f : M \to R^n \times R^m \times \ldots \times R^m = R^n \times (R^m)^n$ by the equation

$$f(x) = (\Phi(x), \varphi_1(x) \cdot h_1(x), \ldots, \varphi_n(x) \cdot h_n(x)) \quad .$$

Of course, $\varphi_i(x) \cdot h_i(x)$ is defined only for x in U_i, but if it is defined to be zero outside U_i, it becomes a C^∞ function on all of M; this we assume done.

f is clearly of class C^r. To prove it is 1-1, suppose $f(x) = f(y)$. Then $\varphi_i(x) = \varphi_i(y)$ for every i, so that if x belongs to V_i, y must also. From the fact that $\varphi_i(x) \cdot h_i(x) = \varphi_i(y) \cdot h_i(y)$ and $\varphi_i(x) = \varphi_i(y) > 0$, it follows that $h_i(x) = h_i(y)$, contradicting the fact that h_i is 1-1.

Since M is compact, f is a homeomorphism. We must show, finally, that it has rank m. Let x_0 belong to W_k, and let us use the coordinate system (U_k, h_k) for computing the rank of f at x_0. We need to show that the matrix $D(f\,h_k^{-1})$ contains a non-singular $m \times m$ matrix. Let $h_k(x) = (u^1, \ldots, u^m) = u$ and $h_k(x_0) = u_0$. Now $\varphi_k(x) \cdot h_k(x) = h_k(x)$ for x in a neighborhood of x_0, so that the $m \times m$ submatrix $\partial(\varphi_k(h_k^{-1}(u)) \cdot u)/\partial u$ of $D(fh_k^{-1})$ is the identity matrix when $u = u_0$.

<u>Remark</u>. The proof just given generalizes to the non-compact case, but there are a few additional difficulties. One arises from the difficulty of finding finitely many coordinate systems which cover M. The other arises from the fact that f need not be a homeomorphism just because it is 1-1. These difficulties are disposed of in the following lemmas and problems (2.6 - 2.9); we then leave the proof of the general theorem to the student (2.10).

2.5 Definition. If φ is a non-negative function defined on M, the carrier of φ is the closure of the set of points x for which $\varphi(x) > 0$. A C^r partition of unity on M is a collection of non-negative functions φ_i of class C^r defined on M, such that the sets C_i = carrier φ_i form a locally-finite covering of M, and such that $\Sigma_i \varphi_i(x) = 1$ for each x. The same definition holds for a topological space, except differentiability has no meaning.

2.6 Problem. Let $\{U_i\}$ be a locally-finite open covering of the C^r manifold M. There is a C^r partition of unity $\{\varphi_i\}$ such that carrier $\varphi_i \subset U_i$ for each i; $\{\varphi_i\}$ is said to be dominated by the covering $\{U_i\}$. In the case $r = 0$, this holds for a space X (separable and metric, as always). Hint: Use 1.3 and 1.4.

(a) Exercise. Let $\{C_i\}$ be a locally-finite covering of the separable metric space X by compact sets; let ϵ_i be a sequence of positive constants. There is a positive real-valued function $\delta(x)$ defined on X such that $\delta(x) < \epsilon_i$ for x in C_i.

2.7 Lemma. Let M be an m-manifold; let $\mathfrak{A}$ be an open covering of M. There is a locally-finite open refinement of $\mathfrak{A}$ which is the union of m+1 collections $\mathfrak{B}_0, \ldots, \mathfrak{B}_m$ such that the sets belonging to any one $\mathfrak{B}_i$ are pairwise disjoint.

Proof. We apply a fundamental theorem of dimension theory: If M is an m-manifold, then every open covering $\mathfrak{A}$ of M has a refinement $\mathfrak{C}$ such that no point of M lies in more than m+1 elements of $\mathfrak{C}$. A proof for the case M compact may be found in [3, p. 67]. The result may readily be extended to the non-compact case, but it is just as easy to prove the theorem from scratch. This is done in 2.12-2.15.

Let $\mathfrak{C} = \{U_j\}$; let $\{V_j\}$ be a locally-finite open covering such that $V_j \subset U_j$ for each j; let $\{\varphi_j\}$ be a partition of unity dominated by $\{V_j\}$, by 1.4 and 2.6. We now may assume the indices are positive integers Given an integer n such that $0 \leq n \leq m$, consider any set $i_0, \ldots, i_n$

of distinct positive integers. Let $W(i_0, \ldots, i_n)$ be the open set consisting of those x for which

$$\varphi_i(x) < \min(\varphi_{i_0}(x), \ldots, \varphi_{i_n}(x))$$

for all values of i other than $i_0,\ldots,i_n$. The collection $\mathcal{B}_n$ is to consist of all such sets $W(i_0,\ldots,i_n)$.

We must prove the elements of $\mathcal{B}_n$ to be pairwise disjoint. Let $i_0,\ldots,i_n$ and $j_0,\ldots,j_n$ be different sets of indices; suppose k belongs to the first set and not to the second, and ℓ belongs to the second set and not to the first. If x is in $W(i_0,\ldots,i_n)$, then $\varphi_\ell(x) < \varphi_k(x)$; if x is in $W(j_0,\ldots,j_n)$, then $\varphi_k(x) < \varphi_\ell(x)$. Hence these two sets are disjoint, as desired.

We prove that $\mathcal{B} = \mathcal{B}_0 \cup \ldots \cup \mathcal{B}_m$ covers M. If x is in M, let $i_0,\ldots,i_n$ be those indices for which $\varphi_i(x) > 0$. There is some such index, since $\Sigma\, \varphi_i(x) = 1$; there are at most $m+1$ of them, by choice of the covering U_j. Then $\varphi_i(x) = 0$ for all other indices i, so that x is in $W(i_0,\ldots,i_n)$.

The fact that $\mathcal{B}$ is locally-finite follows similarly. Each point x has a neighborhood U on which all but a finite number of the functions φ_i are identically zero. Let $i_0, \ldots, i_{\hat{n}}$ be those indices for which φ_i is not identically zero on U. The only sets $W(j_0, \ldots, j_k)$ which U can intersect are those for which $\{j_0, \ldots, j_k\}$ is a subset of $\{i_0, \ldots, i_{\hat{n}}\}$; these are finite in number.

(a) Exercise*. Let M be a simplicial complex in the plane, with vertices $v_1,v_2,\ldots$. Let V_i be the (open) star of v_i, and let $\varphi_i(x)$ be the barycentric coordinate of x with respect to v_i. (It is zero unless x is in V_i, of course.) Carry out the preceding construction in this case, and sketch the resulting sets of the form $W(i_0)$ and $W(i_0,i_1)$.

This exercise should make clearer the idea of the proof of the preceding theorem.

<u>2.8 Problem</u>. Let M be a C^r manifold of dimension m. Prove that M may be covered by $m+1$ coordinate systems $\{(U_i,h_i)\}$. Indeed,

choose U_i so that it is the union of a countable collection of disjoint open sets V_{ij}, where $h_i(V_{ij})$ is a bounded subset of R^m.

2.9 Definition. Let X and Y be separable metric spaces; let $f : X \to Y$ be continuous. The limit set of f is defined as the set of all y in Y such that for some sequence x_n in X having no convergent subsequence, $f(x_n)$ converges to y. It is denoted by $L(f)$.

(a) Exercise. $L(f)$ is a closed subset of Y, if X is locally compact.

(b) Exercise. Let $f : X \to Y$ be continuous and 1-1. $L(f) \cap f(X)$ is empty if and only if f is a homeomorphism; $f(X)$ is closed in Y if and only if $L(f) \subset f(X)$.

2.10 Problem. Let M be a C^r manifold of dimension m. Prove that M has a C^r imbedding as a closed subset of euclidean space of dimension $(m + 1)^2$.

Remark. This theorem shows that any differentiable manifold is a submanifold of some euclidean space. This fact will be of crucial importance to us in the next chapter.

A natural question to ask now is whether one needs to use a euclidean space of such a high dimension in order to imbed M. The answer is no. It is a classic result of Whitney that R^{2m+1} will suffice [15]. Milnor's notes contain a neat proof of this. In fact, R^{2m} will suffice, but the difficulties involved in proving this are much greater.

2.11 Problem*. Let M be a C^r manifold of dimension m. Show that if A is a closed subset of M and h_1 is a C^r imbedding of a neighborhood U of A in R^p, then there is a C^r imbedding f of M in some euclidean space which agrees with h_1 on A. $f(M)$ may be chosen to be closed if $h_1(A)$ is.

Hint: Cover $M' - A$ by coordinate systems (U_i, h_i) for $i = 2, \ldots, m+2$. Let $U_1 = U$; let $\varphi_1, \ldots, \varphi_{m+2}$ be a C^r partition of unity dominated by the covering $U_1, \ldots, U_{m+2}$. Then proceed as in 2.4.

2.12 Definition. A space A has covering dimension at most m if every open covering of A has a refinement of order at most m (which means that no point of A lies in more than $m+1$ elements of the refinement). If A is a closed subset of X, this is equivalent to the statement that every open covering of X has a refinement whose restriction to A has order at most m. If A' is a closed subset of A and A has covering dimension at most m, so does A'.

2.13 Lemma. Let A_n be a closed subset of X, such that $A_n \subset \operatorname{Int} A_{n+1}$ and $\cup A_n = X$. If A_1 and each set $\mathcal{C}\ell(A_{n+1} - A_n)$ have covering dimension at most m, so does X.

Proof. Given an open covering of X, let $\mathcal{B}_0$ be a refinement of it such that any element of $\mathcal{B}_0$ which intersects A_i lies in A_{i+1}. Let $\mathcal{B}_1$ be a refinement of $\mathcal{B}_0$ whose restriction to A_1 has order at most m. Let A_0 be empty for convenience.

Assume that $\mathcal{B}_n$ is a refinement of $\mathcal{B}_0$ whose restriction to A_n has order at most m. Let $\mathcal{C}$ be a refinement of $\mathcal{B}_n$ whose restriction to $\mathcal{C}\ell(A_{n+1} - A_n)$ has order at most m. Define $\mathcal{B}_{n+1}$ as follows:

(1) If U is in $\mathcal{B}_n$ and intersects A_{n-1}, let U be in $\mathcal{B}_{n+1}$.

(2) For each V in $\mathcal{C}$ which intersects A_n but not A_{n-1}, choose an element $f(V)$ in $\mathcal{B}_n$ which contains V. Then for each U in $\mathcal{B}_n$, let U' denote the union of those V in $\mathcal{C}$ for which $f(V)$ is defined and equals U. Let U' belong to $\mathcal{B}_{n+1}$.

(3) For each V in $\mathcal{C}$ which does not intersect A_n, let V belong to $\mathcal{B}_{n+1}$.

It is easy to check that $\mathcal{B}_{n+1}$ covers X and that its restriction to A_{n+1} has order at most m. Let $\mathcal{B}$ consist of those open sets which belong to $\mathcal{B}_n$ for all but finitely many n. Then $\mathcal{B}$ has order at most m.

<u>2.14 Corollary</u>. If B_1 and B_2 are closed subsets of Y which have covering dimension at most m, so does $B_1 \cup B_2$.

<u>Proof</u>. Let $A_1 = B_1$ and let $A_n = X = B_1 \cup B_2$ for $n \geq 2$; apply the preceding lemma. Since $\mathcal{C}\ell(A_2 - A_1)$ is contained in A_2, it has covering dimension at most m.

<u>2.15 Theorem</u>. Every m-manifold M has covering dimension at most m.

<u>Proof</u>. (1) Any compact subset B of R^m has covering dimension at most m: B lies in some cube C, which in turn is the polyhedron of a simplicial complex K of dimension m. Given any open covering $\mathfrak{a}$ of C, there is a subdivision K' of K sufficiently fine that the collection of stars of the vertices of K' are a refinement of $\mathfrak{a}$. This refinement has order m.

(2) Cover M by a locally-finite collection of sets B_i which are homeomorphs of compact subsets of R^m. Let $A_1 = B_1$. In general, given $A_n = B_1 \cup \ldots \cup B_q$, choose an integer $p > q$ such that $A_n \subset \text{Int}(B_1 \cup \ldots \cup B_p)$, and let $A_{n+1} = B_1 \cup \ldots \cup B_p$. Each A_n has covering dimension at most m, by 2.14 The theorem follows from 2.13.

§3. Mappings and Approximations.

In this section we define what is meant by a strong C^1 approximation to a differentiable map of one manifold into another (3.5); in order to do this neatly, it is necessary first to construct the tangent bundle of a differentiable manifold. We then prove (3.10) an important theorem about approximations, which states that if $f : M \to N$ is an immersion, imbedding, or diffeomorphism, then so is any sufficiently good strong C^1 approximation g to f. (In the case of a diffeomorphism, we need to assume that $g(\text{Bd } M) \subset \text{Bd } N$.)

3.1 Definition. Let M be a C^r manifold of dimension m; let x be a point of M. A tangent vector $\vec{v}$ to M at x_0 is a correspondence assigning to every coordinate system (U, h) about x_0 a column matrix α of size $m \times 1$ with the following condition:

If (U, h) and (V, k) are two coordinate systems about x_0, and if α and $\bar{\alpha}$ are the corresponding column matrices, then

$$\bar{\alpha} = D(kh^{-1})(u_0) \cdot \alpha ,$$

where $u_0 = h(x_0)$. The entries of α are called the components of $\vec{v}$ in the coordinate system (U, h).

If $f : [a, b] \to M$ is a C^r map, where $[a, b]$ is an interval in R^1, f is called a (parametrized) curve in M. Let $f(t_0) = x_0$. If we define $\vec{v}$ as assigning to the coordinate system (U, h) the matrix $D(hf)(t_0)$, then $\vec{v}$ is a tangent vector to M at x_0, as the reader may verify. It is called the velocity vector of the parametrized curve f.

We should note that every tangent vector $\vec{v}$ to M at x_0 is the velocity vector of some parametrized curve: Let $\alpha^1, \ldots, \alpha^m$ be the components of $\vec{v}$ in the coordinate system (U, h); let $h(x_0) = 0$. Define $f(t) = h^{-1}(\alpha^1 \cdot t, \ldots, \alpha^m \cdot t)$. If x_0 is in Int M, f is well-defined on some interval $[-\epsilon, \epsilon]$ about 0; if x_0 is in Bd M, f will either be well-defined on $[-\epsilon, 0]$ or $[0, \epsilon]$. In either case, $\vec{v}$ will be the velocity vector of f at x_0.

Note that the tangent vectors to M at x_0 form an m-dimensional vector space; we merely take the components of $\vec{v}$ and $\vec{w}$ in the coordinate system (U,h) and carry out addition component-wise. One checks that this definition is independent of coordinate system. This space is called the <u>tangent space</u> to M at x_0.

<u>3.2 Definition</u>. Let M be a C^r manifold. Let $T(M)$ denote the set of all tangent vectors to M; let $\pi : T(M) \to M$ carry the vector $\vec{v}$ at x_0 into the point x_0. $T(M)$ is called the <u>tangent bundle</u> of M, and π is called the <u>projection mapping</u>.

If (U, h) is a coordinate system on M, define $\tilde{h} : \pi^{-1}(U) \to R^m \times R^m$ or $H^m \times R^m$ by the equation

$$\tilde{h}(\vec{v}) = h\pi(\vec{v}) \times (\alpha^1, \ldots, \alpha^m)$$

where $(\alpha^1,\ldots,\alpha^m)$ are the components of $\vec{v}$ in the coordinate system (U,h). The requirement that $\pi^{-1}(U)$ be open and that $\tilde{h}$ be a homeomorphism immediately imposes a topology on $T(M)$ which is Hausdorff and separable (see Exercise (a)).

$T(M)$ is a 2m-manifold with $\pi^{-1}(\text{Bd } M)$ as its boundary, for the pairs $(\pi^{-1}(U),\tilde{h})$ will be coordinate neighborhoods on $T(M)$.

If (U,h) and (V,k) are overlapping coordinate neighborhoods on M, then

$$\tilde{k}\,\tilde{h}^{-1}(x, \alpha) = (kh^{-1}(x), D(kh^{-1})(x) \cdot \alpha)$$

on $h(U \cap V) \times R^m$, where α is written as a column vector. Thus $\tilde{k}\,\tilde{h}^{-1}$ is of class C^{r-1}, so that the collection of pairs $(\pi^{-1}(U), \tilde{h})$ serves as a basis for a differentiable structure of class C^{r-1} on $T(M)$. Whenever we consider $T(M)$, we will assume it provided with this differentiable structure, if $r > 1$.

The <u>inclusion mapping</u> $i : M \to T(M)$ which carries x into the zero vector at x, and the projection mapping π, are maps of class C^{r-1}.

(a) Exercise. Let (U,h) and (V,k) be two overlapping coordinate systems on M. Show that $\tilde{k}\,\tilde{h}^{-1}$ is a homeomorphism of $h(U \cap V) \times R^m$ with $k(U \cap V) \times R^m$. Conclude that the topology on $T(M)$ is well-defined.

Show also that it is separable and Hausdorff.

(b) Exercise*. Show that for some neighborhood V of the point x of M, there is a diffeomorphism $g : \pi^{-1}(V) \to V \times R^m$ such that πg^{-1} is the natural projection of $V \times R^m \to V$, and g is a linear isomorphism of the tangent space at x onto $x \times R^m$.

This shows that $\pi : T(M) \to M$ is a fibre bundle, with R^m as fibre and the non-singular linear transformations of R^m as structural group. (See [12].)

3.3 Definition. Let M and N be manifolds of class at least C^r; let $g : M \to N$ be a C^r map. There is an induced map $dg : T(M) \to T(N)$ of class C^{r-1}, defined as follows: Let $\vec{v}$ be a vector at x_0 assigning the matrix α to the coordinate system (U, h), and let $\vec{w}$ be a vector at $g(x_0)$ assigning β to the coordinate system (V, k). We define $dg(\vec{v}) = \vec{w}$ if

$$\beta = D(kgh^{-1}) \cdot \alpha \quad .$$

Note that dg is a linear map on each tangent space; it is called the differential of g.

(a) Exercise. Check that dg is well-defined, independently of coordinate system.

(b) Exercise. Show that if $f : [a,b] \to M$ is a C^1 curve having $\vec{v}$ as velocity vector, then $dg(\vec{v})$ is the velocity vector of the curve $gf : [a,b] \to N$.

(c) Exercise. If $f : [a,b] \to M$ is a C^1 map, interpret df geometrically, noting that f maps the 1-dimensional manifold $[a,b]$ into M. The preceding exercise then merely states that $d(gf) = dg \cdot df$. Generalize this formula to the case where the domain of f is a manifold of dimension greater than 1.

(d) Exercise*. If g is an immersion, an imbedding, or a diffeomorphism, how is this fact reflected by the map dg ?

3.4 Definition. Let M be a C^r manifold. Suppose we have an inner product $[\vec{v}, \vec{w}]$ defined on each tangent space of M, which is differentiable of class C^q, $0 \leq q < r$, in the sense that the map of $T(M)$ into R which carries $\vec{v}$ into $[\vec{v}, \vec{v}]$ is of class C^q. Such an inner product is called a Riemannian metric on M. As usual, $\|\vec{v}\|$ is used to denote $\sqrt{[\vec{v}, \vec{v}]}$.

Let us note that it follows from the identity

$$[\vec{v}, \vec{w}] = \tfrac{1}{4}\|\vec{v} + \vec{w}\|^2 - \tfrac{1}{4}\|\vec{v} - \vec{w}\|^2$$

that the real function $[\vec{v}, \vec{w}]$ is continuous on the subspace of $T(M) \times T(M)$ consisting of all pairs $(\vec{v}, \vec{w})$ such that $\pi(\vec{v}) = \pi(\vec{w})$.

(a) Exercise. There is a standard Riemannian metric on the manifold R^n; we simply take the components α^i and β^i of $\vec{v}$ and $\vec{w}$ relative to the coordinate system $i : R^n \to R^n$ and define $[\vec{v},\vec{w}]$ as their dot product, $\Sigma\, \alpha^i \beta^i$.

Let M be a C^r manifold; let $f : M \to R^n$ be a C^{q+1} immersion. If $\vec{v}$ and $\vec{w}$ are tangent to M at x, let us define $[\vec{v},\vec{w}]$ to equal the dot product in $T(R^n)$ of $df(\vec{v})$ and $df(\vec{w})$. Check that this is a Remannian metric on M. Why do we need f to be an immersion?

(b) Exercise.* Let $[\vec{v},\vec{w}]$ be a Riemannian metric on M; let (U, h) be a coordinate system on M, and let α and β be the matrices corresponding to $\vec{v}$ and $\vec{w}$ under this coordinate system. Show there is a unique C^q matrix function $G(x)$ defined on U which is symmetric and positive definite, such that

$$[\vec{v},\vec{w}] = \beta^{tr} \cdot G(x) \cdot \alpha \quad .$$

Let $\bar{G}(x)$ be the matrix function corresponding to the coordinate system (V, k); show that

$$\bar{G}(x) = D(hk^{-1})^{tr} \cdot G(x) \cdot D(hk^{-1}) \quad .$$

Any correspondence assigning to each coordinate system (U, h) about x_0, a matrix G, with this equation relating G and $\bar{G}$, is called classically a covariant tensor of second order.

3.5 Definition. If $\vec{v}$ is a tangent vector to R^n at x, let α^i be the components of $\vec{v}$ relative to the coordinate system $i : R^n \to R^n$.

The correspondence $\vec{v} \to (x, \alpha)$ gives a homeomorphism between $T(R^n)$ and $R^n \times R^n$ which is an isomorphism of the tangent space at x onto $x \times R^n$. We often identify $T(R^n)$ with $R^n \times R^n$ for convenience, and we write $\vec{v} = (x, \alpha)$. If $f : M \to R^n$ is a C^1 map and $\vec{v}$ is a tangent vector to M at x, we define $d_0 f(\vec{v})$ by the equation $df(\vec{v}) = (f(x), d_0 f(\vec{v}))$.

Consider $F^1(M, R^n)$, the space of all C^1 maps of M into R^n. We topologize this space as follows:

Choose a Riemannian metric on M. Given the C^1 map $f : M \to R^n$, and the positive continuous function $\delta(x)$ on M, let $W(f, \delta)$ denote the set of all C^1 maps $g : M \to R^n$ such that

$$\|f(x) - g(x)\| < \delta(x)$$

and

$$\|d_0 f(\vec{v}) - d_0 g(\vec{v})\| < \delta(x) \|\vec{v}\|$$

for each x in M and each $\vec{v} \neq \vec{0}$ tangent to M at x. If g is in $W(f, \delta)$, then g is called a δ-approximation to f. It is also called a strong C^1 approximation to f. The sets $W(f, \delta)$ form a basis for what is called the fine C^1 topology on $F^1(M, R^n)$.

Given a differentiable manifold N, let us imbed it differentiably in some R^n. The fine C^1 topology on $F^1(M, N)$ is defined to be that induced from the fine C^1 topology on $F^1(M, R^n)$. This topology is independent of the Riemannian metric on M and the choice of the imbedding of N (see Exercise (b) of 3.6).

To obtain the coarse C^1 topology, one alters the definition to require that the inequalities hold only for x in some compact set A; one takes sets $W(f, \delta, A)$ as a basis. We shall make no essential use of the coarse C^1 topology.

(a) Exercise. Check that the sets $W(f,\delta)$ form a basis for a topology.

(b) Exercise. The fine C^0 topology on $F^0(X,Y)$ (the space of all continuous maps of the separable metric space X into the separable metric space Y) is obtained as follows: Given $f : X \to Y$ and a positive continuous function δ on X, let a neighborhood of f consist of all maps g such that $\rho(f(x), g(x)) < \delta(x)$ for all x, where ρ is a (topological) metric on Y. The map g is called a strong C^0 approximation to f, or simply, a C^0 δ-approximation to f. Check that these sets form a basis for a topology.

3.6 Lemma. Let N be a submanifold of R^n; let $f : M^m \to N^p$ be a C^1 map. Let (U, h) be a coordinate system on M; let C be a compact subset of U; let (V, k) be a coordinate system on N about $f(C)$. Given $\epsilon > 0$, there exists a $\delta > 0$ such that if $g : M \to N$ and

$$(*) \qquad g(C) \subset V,$$
$$|kfh^{-1}(y) - kgh^{-1}(y)| < \delta,$$
$$|D(kfh^{-1})(y) - D(kgh^{-1})(y)| < \delta,$$

for all $y \in h(C)$, then

$$(**) \qquad \|f(x) - g(x)\| < \epsilon,$$
$$\|d_0 f(\vec{v}) - d_0 g(\vec{v})\| < \epsilon \|\vec{v}\|,$$

for each non-zero $\vec{v}$ tangent to M at a point x of C.

Proof. Suppose not. Then there exists for each n a function g_n and a vector $\vec{v}_n \neq \vec{0}$ tangent to M at a point x_n of C, such that the inequalities in (*) hold for $g = g_n$, $y = y_n = h(x_n)$, and $\delta = 1/n$, but the inequalities in (**) do not hold for $g = g_n$, $x = x_n$, and $\vec{v} = \vec{v}_n$, and all large n. By passing to subsequences and renumbering, we may assume $x_n \to x$ and $\vec{v}_n/\|\vec{v}_n\| = \vec{u}_n \to \vec{u}$, a unit tangent to M at x.

Now $|kf(x_n) - kg_n(x_n)| < 1/n$, so that $kf(x_n)$ and $kg_n(x_n)$ converge to $kf(x)$. Then both $f(x_n)$ and $g_n(x_n)$ converge to $f(x)$, so that $\|f(x_n) - g_n(x_n)\| \to 0$.

Let $dh(\vec{u}_n) = (y_n, \alpha_n)$, in $R^m \times R^m = T(R^m)$. Then

$$|D(kfh^{-1})(y_n) \cdot \alpha_n - D(kg_n h^{-1})(y_n) \cdot \alpha_n| < m|\alpha_n|/n .$$
$$\|d_0 k(df(\vec{u}_n)) - d_0 k(dg_n(\vec{u}_n))\| < \sqrt{p}\, m\|dh(\vec{u}_n)\|/n .$$

Since the right side approaches zero, so does the left. Hence both $d_0 k(df(\vec{u}_n))$ and $d_0 k(dg_n(\vec{u}_n))$ converge to $d_0 k(df(\vec{u}))$, so that $dk(df(\vec{u}_n))$ and $dk(dg_n(\vec{u}_n))$ converge to $dk(df(\vec{u}))$. Since $d(k^{-1})$ is continuous, $df(\vec{u}_n)$ and $dg_n(\vec{u}_n)$ converge to $df(\vec{u})$, so $\|d_0 f(\vec{u}_n) - d_0 g_n(\vec{u}_n)\| \to 0$. Contradiction.

(a) Exercise. Prove the converse of this lemma: Given $\delta > 0$, there is an $\epsilon > 0$ such that if (**) holds, so does (*).

(b) Exercise. Let $f : M \to N$. Let $\{C_i\}$ be a locally finite covering of M by compact sets; let (U_i, h_i) and (V_i, k_i) be coordinate systems about C_i and $f(C_i)$ respectively. Let $X(f, \delta_i)$ consist of all maps $g : M \to N$ such that $g(C_i) \subset U_i$, $|k_i f h_i^{-1}(y) - k_i g h_i^{-1}(y)| < \delta_i$, and $|D(k_i f h_i^{-1})(y) - D(k_i g h_i^{-1})(y)| < \delta_i$, for all y in $h_i(C_i)$. Show this neighborhood system of f is equivalent to the fine C^1 neighborhood system of f. Conclude that the fine C^1 topology on $F^1(M, N)$ is independent of the Riemannian metric on M and the imbedding of N.

(c) Exercise. Let $X(f, \delta_i)$ be defined as in the preceding exercise, except that the second inequality (involving the derivatives of f) is deleted. Show the resulting neighborhood system of f is equivalent to the fine C^0 neighborhood system of f. Conclude that the fine C^0 topology on $F^0(M, N)$ is independent of the metric chosen for N.

(d) Exercise*. Show that the topology on $F^1(M, N)$ induced from the fine C^0 topology is strictly coarser than the fine C^1 topology.

(e) Exercise*. The coarse C^1 topology on $F^1(M, N)$ is defined above. Formulate an alternate definition similar to that given in the statement of Lemma 3.6, and prove the two definitions equivalent.

(f) Exercise*. Consider $F^1(R, R)$ in the fine and coarse C^1 topologies, and in the uniform topology, whose basis elements are of the form $W(f, \epsilon)$, where $f : R \to R$, $\epsilon > 0$, and $W(f, \epsilon)$ is the set of all g such that $\text{lub}\, |f(x) - g(x)| < \epsilon$. Describe the path-component of the zero function in each case.

<u>3.7 Problem</u>. Let $f : M \to N$ and $g : N \to P$ be C^1 maps.

(1) Given $\epsilon(x) > 0$, defined on M, prove there is a $\delta(x) > 0$ such that whenever $\tilde{f}$ is a δ-approximation to f, then $g\tilde{f}$ is an ϵ-approximation to gf.

(2) If f imbeds M as a closed subset of N, prove there exist $\delta_1(x) > 0$ and $\delta_2(y) > 0$ such that whenever $\tilde{f}$ is a δ_1-approximation to f and $\tilde{g}$ is a δ_2-approximation to g, then $\tilde{g}\tilde{f}$ is an ϵ-approximation to gf.

(3) If f is a diffeomorphism, show that given $\epsilon(y) > 0$, there exists $\delta(x) > 0$ such that if $\tilde{f}$ is a δ-approximation to f and is a diffeomorphism, then $\tilde{f}^{-1}$ is an ϵ-approximation to f^{-1}.

(a) Exercise.* Show that (2) may fail if f is not a homeomorphism, or if $f(M)$ is not closed.

3.8 Definition. Consider the set $F^r(M,N)$ of C^r maps of M into N. The definition of the fine C^1 topology given in Lemma 3.6 generalizes to defining the fine C^r topology for this space. The same definition is used, except that the second equality (which requires closeness of the first partial derivatives) is replaced by the statement that the partial derivatives of $k_i g h_i^{-1}$, through order r, approximate the corresponding partial derivatives of $k_i f h_i^{-1}$ within δ_i, for x in C_i. The fine C^∞ topology on $F^\infty(M, N)$ is the one having as basis the union of the C^r topologies for all finite r. We shall have no occasion to use these topologies.

3.9 Definition. Let f_0 and f_1 be C^1 maps of M into N. A regular C^1 homotopy between f_0 and f_1 is a continuous map $f_t\colon M \times R \to N$ such that

(1) For some $\epsilon > 0$, $f_t = f_0$ for $t < \epsilon$ and $f_t = f_1$ for $t > 1 - \epsilon$.

(2) f_t is a C^1 map for each t.

(3) $df_t\colon T(M) \times R \to T(N)$ is continuous.

The homotopy f_t is said to be a differentiable C^1 homotopy if the map $f_t\colon M \times R \to N$ is of class C^1; this is stronger than conditions (2) and (3). It is a differentiable C^r homotopy if it is of class C^r.

If f_0 and f_1 are both immersions, we make the (standard) convention that a "regular (or differentiable) homotopy f_t" between them always means that f_t is an immersion for each t. If f_0 and f_1 are imbeddings, a regular (or differentiable) homotopy f_t is said to be an isotopy if f_t is an imbedding for each t.

The two notions of regular and differentiable homotopies in fact differ only slightly. It will appear later as a problem, to prove that the existence of a regular homotopy between f_0 and f_1 implies the existence of a differentiable one. It is not clear, however, which notion should be given preference as being more natural. A regular homotopy is natural in that it is equivalent to the existence of a path $\Phi : R \to F^1(M,N)$ between the maps f_0 and f_1, where the coarse C^1 topology is used for the function space. However, the homotopies one constructs in practice are usually differentiable rather than merely regular. This is the case with us, so we shall have little occasion to use the weaker notion.

(a) Exercise. Let f_t be a homotopy between f_0 and f_1; let g_t be a homotopy between g_0 and g_1; let $f_1 = g_0$. Define $h_t : M \times R \to N$ by the equations

$$h_t(x) = f_{2t}(x) \qquad \text{for} \quad t \leq 1/2$$
$$h_t(x) = g_{2t-1}(x) \qquad \text{for} \quad t \geq 1/2 \quad .$$

Show that h_t is a regular (or differentiable) C^1 homotopy if f_t and g_t are.

(b) Exercise*. Construct a regular C^1 homotopy which is not a differentiable C^1 homotopy.

(c) Exercise*. Let f_t be a regular C^1 homotopy. Define $\Phi : R \to F^1(M,N)$ by the equation

$$\Phi(t)(x) = f_t(x) \quad .$$

Show that Φ is continuous, if the coarse C^1 topology is used for $F^1(M, N)$. State and prove the converse.

<u>3.10 Theorem</u>. Let $f : M \to N$ be a C^1 map. If f is an immersion or imbedding, there is a fine C^1 neighborhood of f consisting only of immersions or imbeddings, respectively. If f is a diffeomorphism, there is a fine C^1 neighborhood of f such that if g is in this neighborhood and g carries Bd M into Bd N, then g is a diffeomorphism.

<u>Proof</u>. Let $\{C_i\}$ be a locally-finite covering of M by compact sets, such that the sets Int C_i also cover M, and (U_i,h_i) and (V_i,k_i) are coordinate systems about C_i and $f(C_i)$, respectively.

(1) Suppose f is an immersion. Then the matrix $D(k_i f h_i^{-1})(x)$ has rank $m = \dim M$ for all x in $h_i(C_i)$. We denote the set of $p \times q$ matrices of rank r by $M(p,q;r)$; it is considered as a subspace of R^{pq}. Now the set $M(m,n;m)$ is open in R^{mn}: If one maps each matrix into the sum of the squares of the determinants of its $m \times m$ submatrices, the set is the inverse image of the open set $R - \{0\}$ under this continuous map.

The matrices $D(k_i f h_i^{-1})(x)$, where x is in $h_i(C_i)$, form a compact subset K_i of $M(m,n;m)$. Hence there is a δ_i such that the δ_i-neighborhood of K_i lies in $M(m,n;m)$. This choice of the δ_i specifies the desired fine C^1 neighborhood of f, as in 3.6.

(2) Let f be an imbedding. We have just proved that there is a fin C^1 neighborhood W_1 of f consisting of immersions. Within this, there i a fine C^1 neighborhood W_2 consisting of maps g which are 1-1 on each

For otherwise, there would exist a number i and a sequence g_n where g_n converges to f, and dg_n converges to df, uniformly on C_i, but $g_n(x_n) = g_n(y_n)$ for two distinct points x_n and y_n of C_i. By passing to subsequences and renumbering, we may assume $x_n \to x$ and $y_n \to y$ Since $f(x_n) \to f(x)$, $g_n(x_n) \to f(x)$; similarly, $g_n(y_n) \to f(y)$. Hence $f(x) = f(y)$; since f is 1-1, $x = y$.

We refer f and g to coordinate systems: let $\bar{f} = k_i f h_i^{-1}$, $\bar{g} = k_i g h_i^{-1}$, $\bar{x}_n = h_i(x_n)$, and $\bar{y}_n = h_i(y_n)$. Then

$$0 = \bar{g}_n^i(\bar{x}_n) - \bar{g}_n^i(\bar{y}_n) = D\bar{g}_n^i(z_{n,i}) \cdot (\bar{x}_n - \bar{y}_n) ,$$

for some point $z_{n,i}$ on the line segment joining $\bar{x}_n$ and $\bar{y}_n$. By passing to a subsequence and renumbering, we may assume $(\bar{x}_n - \bar{y}_n)/\|\bar{x}_n - \bar{y}_n\| \to \vec{u}$. Then since $z_{n,i} \to \bar{x}$, we have

$$0 = D\bar{f}^i(\bar{x}) \cdot \vec{u} ,$$

for each i, contradicting the fact that $D\bar{f}(\bar{x})$ is non-singular.

(3) Let f be an imbedding. We have just proved there is a fine C^1 neighborhood W_2 of f consisting of immersions which are 1-1 on each C_i. Now we prove there is a fine C^0 neighborhood of f whose intersection with W_2 consists of maps which are globally 1-1; we shall denote this intersection by W_3. We specify this neighborhood by a continuous function $\delta(x)$; thus we assume N is given a topological metric ρ. There is a covering D_i of M by compact sets, with $D_i \subset \text{Int}\, C_i$ for each i (see 1.4). Let ϵ_i be the distance in N from $f(D_i)$ to $f(M - \text{Int}\, C_i)$; since f is a homeomorphism, this distance is positive. Let $\delta(x)$ be a continuous function on M which is less than $\epsilon_i/2$ on C_i (see 2.6). Let g be a C^0 δ-approximation to f which lies in W_2. Suppose $g(x) = g(y)$, where

x is in D_i and y is in D_j, and $\epsilon_i \leq \epsilon_j$. Then

$$\rho(f(x), f(y)) < \epsilon_i/2 + \epsilon_j/2 \leq \epsilon_j.$$

Since g is 1-1 on C_j, x is not in C_j, so that $\rho(f(x), f(y)) \geq \epsilon_j$ by choice of ϵ_j. Thus we have a contradiction.

Finally, we prove there exists a fine C^0 neighborhood of f whose intersection with W_3 consists of homeomorphisms. Since f is a homeomorphism, $L(f) \cap f(M)$ is empty. Let $\epsilon_i < 1/i$ and let ϵ_i be less than the distance in N from the compact set $f(C_i)$ to the closed set $L(f)$. Let $\delta(x)$ be a continuous function on M such that $\delta(x) < \epsilon_i$ for x in C_i. Let g be a C^0 δ-approximation to f which lies in W_3.

First we show that $L(f) = L(g)$. If x_n is a sequence in M having no convergent subsequence, then each C_i contains only finitely many terms of the sequence. This means that $\rho(f(x_n), g(x_n)) \to 0$, since $\rho(f(x_n), g(x_n)) < 1/i$ if x_n is in C_i. Hence $f(x_n) \to y$ if and only if $g(x_n) \to y$, so that $L(f) = L(g)$.

It follows that $L(g) \cap g(M)$ is empty. If x is in C_i, then $\rho(g(x), f(x)) < \epsilon_i$, so that $g(x)$ is not in $L(f)$, which equals $L(g)$. Hence g is a homeomorphism. Note that $g(M)$ is a closed subset of N if and only if $f(M)$ is.

(4) Let f be a diffeomorphism. There is a fine C^1 neighborhood W_4 of f consisting of imbeddings. The following lemma shows that there is a fine C^0 neighborhood of f such that if g lies in the intersection of W_4 and this neighborhood, and $g(\text{Bd } M) \subset \text{Bd } N$, then g is a diffeomorphism.

(a) Exercise. Let f be a homeomorphism of X into Y; let $\{C_i\}$ be a locally-finite covering of X by compact sets such that the sets Int C_i cover X. Prove there is a fine C^0 neighborhood of f such that if g lies in this neighborhood and g is 1-1 on each C_i, then g is a homeomorphism. (X and Y are separable metric, as always.)

(b) Exercise*. Let M and N be C^1 manifolds; let f_t be a regular C^1 homotopy between the maps f_0 and f_1 of M into N. If f_t is an immersion for each t, prove that for some $\delta(x) > 0$, any C^1 regular homotopy F_t between f_0 and f_1 which is a δ-approximation to f_t for each t, is also an immersion for each t. Prove similar theorems when

f_t is an imbedding and M is compact, or when f_t is a diffeomorphism between non-bounded manifolds.

Hint: Generalize the proof of 3.10. (1) and (2) suffice for the case of an immersion or an imbedding; for the case of a diffeomorphism, (3) and (4) may be applied to the map $F : M \times R \to N \times R$ defined by the equation $F(x, t) = (f_t(x), t)$.

(c) Exercise*. Show the hypothesis that M is compact is needed in the preceding exercise when f_t is an imbedding.

<u>3.11 Lemma</u>. Let f be a homeomorphism of M onto N, where M and N are topological manifolds. There is a fine C^0 neighborhood of f such that if g lies in this neighborhood and $g(\text{Bd } M) \subset \text{Bd } N$, then g carries M onto N.

<u>Proof</u>. We consider first the case in which $\text{Bd } M$ is empty. Choose a locally-finite covering of M by sets C_i such that for some coordinate system (V_i, k_i) about $f(C_i)$, $k_i f(C_i)$ equals the unit m-ball B^m, and the sets $\text{Int } C_i$ also cover M (see Exercise (a)).

Choose δ_i small enough that the ball $B(1 + \delta_i)$ of radius $1 + \delta_i$ is contained in $k_i(V_i)$, and so that the sets $D_i = (k_i f)^{-1}(B(1 - \delta_i))$ still cover M (see 1.4). Then the sets $f(D_i)$ cover N.

Let $g : M \to N$ be a map such that $\| k_i g(x) - k_i f(x) \| < \delta_i$ when x is in C_i, for all i. We prove that g is onto; in particular, we prove that $g(C_i)$ contains $f(D_i)$. Said differently, the composite map $h = (k_i g)(k_i f)^{-1}$ is a well-defined map carrying B^m into R^m, and the image of the unit sphere $S^{m-1} = \text{Bd } B^m$ under h lies outside $B(1 - \delta_i)$; we prove that $h(B^m)$ contains $B(1 - \delta_i)$.

Suppose z lies in $B(1 - \delta_i)$ but not in $h(B^m)$. Let λ be the radial projection from z, of $R^m - z$ onto S^{m-1}. Then λh carries B^m into S^{m-1}.

On the other hand, consider $h|S^{m-1} : S^{m-1} \to R^m$. It is homotopic to the identity; we merely define $F_t(x) = t \cdot h(x) + (1-t) \cdot x$ for x in S^{m-1}. This homotopy carries $h(x)$ along the straight line between $h(x)$ and x, so $F_t(x)$ lies outside $B(1-\delta_i)$. Hence the map λF_t is well-

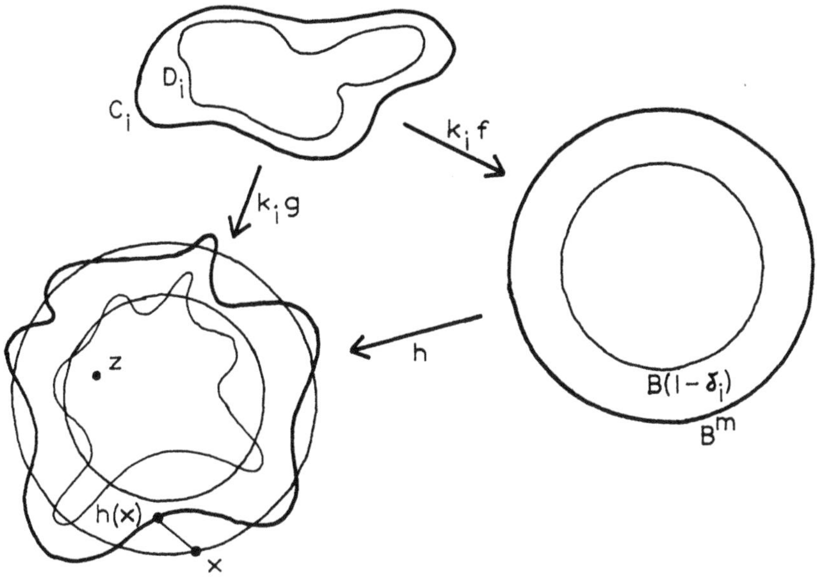

defined, and is a homotopy between $\lambda h|S^{m-1} : S^{m-1} \to S^{m-1}$ and the identity.

Consider the homology sequence of the pair (B^m, S^{m-1}) and the homomorphism of it induced by λh:

$$\begin{array}{ccccccc} 0 \to & H_m(B^m, S^{m-1}) & \to & H_{m-1}(S^{m-1}) & \to 0 \\ & \downarrow (\lambda h)_* & & \downarrow (\lambda h \mid S^{m-1})_* & \\ 0 \to & H_m(B^m, S^{m-1}) & \to & H_{m-1}(S^{m-1}) & \to 0 \end{array}$$

Now $(\lambda h)_*$ is the zero-homomorphism of the infinite cyclic group $H_m(B^m, S^{m-1})$, because λh maps B^m into S^{m-1}. And $(\lambda h \mid S^{m-1})_*$ is the identity homomorphism of the infinite cyclic group $H_{m-1}(S^{m-1})$ because $\lambda h \mid S^{m-1}$ is homotopic to the identity map. This contradicts the commutativity of this diagram. (This argument is the only place where we use a bit of algebraic topology.)

Now let us consider the case in which Bd M is non-empty. Let D(M) be the double of M, as in 1.1; let D(N) be the double of N; let $\tilde{f} : D(M) \to D(N)$ be the homeomorphism induced from f. Choose a metric ρ on D(N). There is a positive continuous function $\epsilon(x)$ on D(M) such that if $\tilde{g} : D(M) \to D(N)$ and $\rho(\tilde{f}(x), \tilde{g}(x)) < \epsilon(x)$, then $\tilde{g}$ is onto.

Now $D(M) = M_0 \cup M_1$; ϵ determines two positive continuous functions ϵ_0 and ϵ_1 on M. Similarly, ρ induces two metrics ρ_0 and ρ_1 on N.

Let $g : M \to N$; if $g(\text{Bd } M) \subset \text{Bd } N$, g induces a map $\tilde{g} : D(M) \to D(N)$. If we require g to be an ϵ_0-approximation to f relative to the metric ρ_0, and an ϵ_1-approximation relative to ρ_1, then $\tilde{g}$ is an ϵ-approximation to $\tilde{f}$ and hence is onto. Then g is necessarily onto as well.

(a) Exercise. Prove the existence of the sets C_i used in the preceding proof.

(b) Exercise. Let A be a closed subset of the non-bounded manifold M; let B be a closed set containing A in its interior. Prove there is a positive continuous function $\delta(x)$ defined on B such that if $f : B \to M$ is a C^0 δ-approximation to the identity, then $f(B)$ contains A.

§4. Smoothing of Maps and Manifolds.

We now approach the two main goals of this chapter. The first of our theorems states that if M and N are C^∞ manifolds, and $f : M \to N$ is a C^1 immersion, imbedding, or diffeomorphism, then f may be approximated by a C^∞ immersion, imbedding, or diffeomorphism, respectively. The proof of this appears in 4.2 - 4.5, except for the case where f is a diffeomorphism and M has a boundary, which is postponed to Section 5.

The second theorem states that every differentiable structure of class C^1 on a manifold M contains a C^∞ structure. The proof in the case where M is non-bounded appears in 4.7 - 4.9; the other case is treat- in Section 5.

The fundamental tool needed for proving these theorems is the following "smoothing lemma."

4.1 Lemma. Let U be an open subset of H^m or R^m. Let A be a compact subset of the open set V, where $\bar{V} \subset U$. Let $f : U \to R^n$ be a C^r map, $1 \leq r$. Let δ be a positive number. There is a map $f_1 : U \to R^n$ such that

(1) f_1 is of class C^∞ in a neighborhood of A.

(2) f_1 equals f outside V.

(3) $|f_1(x) - f(x)| < \delta$ and $|Df_1(x) - Df(x)| < \delta$, for all x.

(4) f_1 is of class C^p on any open set on which f is of class C^p, $1 \leq p \leq \infty$.

(5) There is a differentiable C^r homotopy f_t between $f_0 = f$ and f_1, such that f_t satisfies conditions (2)-(4) for each t.

Proof. It suffices to consider the case in which U is open in R^m, since the other case reduces to this after f is extended to a neighborhood of U in R^m. We may also assume $\bar{V}$ is compact.

Let W be an open set containing A, such that $\bar{W} \subset V$. Let Ψ be a C^∞ function on R^m which equals 1 in a neighborhood of A and equals 0 outside W. (Take a partition of unity dominated by the two-element covering $\{W, R^m - A\}$, and let Ψ be the function corresponding to W.)

Let $g(x) = \Psi(x) \cdot f(x)$; then $g : U \to R^n$. We extend g to R^m by letting it equal zero outside $\bar{W}$; then g is of class C^r.

Let $\varphi(x)$ be a C^∞ function on R^m which is positive on $\text{Int}\, C(\epsilon)$, and zero elsewhere. Here ϵ is a positive number yet to be chosen. We also assume that

$$\int_{C(\epsilon)} \varphi(x)\, dx = 1 \quad ,$$

since this may be obtained by multiplying φ by an appropriate constant. Define

$$h(x) = \int_{C(\epsilon)} \varphi(y)\, g(x + y)\, dy$$

for x in R^m. Choose $\sqrt{m}\epsilon$ less than the distance from W to the complement of V. Then $h(x) = 0$ for x outside V. Define

$$f_1(x) = f(x) \cdot (1 - \Psi(x)) + h(x) \quad .$$

Since $\Psi(x)$ and $h(x) = 0$ for x outside V, condition (2) of the lemma is satisfied.

Now $f_1(x) = h(x)$ for x in a neighborhood of A, and h is a function of class C^∞:

$$\begin{aligned} h(x) &= \int_{C(\epsilon)} \varphi(y)\, g(x + y)\, dy \\ &= \int_{x + C(\epsilon)} \varphi(z - x)\, g(z)\, dz \\ &= \int_{R^m} \varphi(z - x)\, g(z)\, dz \quad , \end{aligned}$$

since $\varphi(z - x) = 0$ if z is not in $x + C(\epsilon)$. Since φ is a C^∞ function, h is clearly of class C^∞. Thus condition (1) holds.

Since $f_1 = f \cdot (1 - \Psi) + h$, and Ψ and h are of class C^∞, the class of f_1 on any open set is no less than the class of f. Hence condition (4) is satisfied.

Now $f_1(x) = f(x) + (h(x) - g(x))$, so that we need merely choose ϵ small enough that h is a δ-approximation to g. By a mean-value theorem,

$$h^i(x) = g^i(x + y_i)$$

and

$$\frac{\partial h^i}{\partial x^j}(x) = \frac{\partial g^i}{\partial x^j}(x + y_{ij})$$

where y_i and y_{ij} are points in $C(\epsilon)$. The functions g^i and $\partial g^i/\partial x^j$

are uniformly continuous, so we need merely choose ϵ so that $|g^i(x) - g^i(x^*)|$ and $|\partial g^i/\partial x^j(x) - \partial g^i/\partial x^j(x^*)|$ are less than δ if $|x - x^*| < \epsilon$; then condition (3) holds.

Finally, let α be a monotonic C^∞ function such that $\alpha(t) = 0$ for $t \leq 1/3$ and $\alpha(t) = 1$ for $t \geq 2/3$ (see 1.3). Define

$$f_t(x) = \alpha(t)\, f_1(x) + (1 - \alpha(t))\, f(x) \quad .$$

Then f_t is a C^r differentiable homotopy between f and f_1. Outside V, $f_1 = f$, so that $f_t = f$ also. Likewise,

$$|f_t - f| = \alpha(t)\,|f_1 - f| < \delta$$

and

$$|Df_t - Df| = \alpha(t)\,|Df_1 - Df| < \delta \quad .$$

Finally, f_t is of class C^p on any open set where f_1 and f are.

(a) Exercise. Consider the proof just given. Show that if U is open in R^m and $f(U) \subset H^n$, then $f_t(U) \subset H^n$. Show that if U is open in H^m and $f(U) \subset H^n$, it need not be true that $f_t(U) \subset H^n$. Modify the construction of f_1 so that this relation will hold. (Hint: Replace $C(\epsilon)$ by $C(\epsilon) \cap H^m$ throughout the proof.)

(b) Exercise. Consider the preceding theorem in the case $r = 0$. Show that the conclusions still hold, except that $|Df_1(x) - Df(x)| < \delta$ no longer makes sense (since Df does not exist), and f_t is of class C^0 and hence only a homotopy.

4.2 Theorem. Let M and N be manifolds of class C^p. Let $f : M \to N$ be a map of class C^r $(1 \leq r \leq p \leq \infty)$. Let $\delta(x) > 0$. There is a C^p map $h : M \to N$ such that

(1) h is a δ-approximation to f.

(2) There is a C^r differentiable homotopy f_t between f and h which is a δ-approximation to f, for each t.

Proof. Let $\{U_i\}$ be a locally-finite covering of M, $\bar{U}_i$ compact, where (U_i, h_i) is a coordinate system on M and $f(U_i)$ is contained in the coordinate system $(\mathcal{O}_i, k_i)$ on N. Let $\{W_i\}$ be an open covering of M, with $\bar{W}_i$ contained in the open set V_i, and $\bar{V}_i$ contained in U_i. Let δ be so small that any δ-approximation to f carries $\bar{U}_i$ into $\mathcal{O}_i$.

Let $f_0 = f$. Assume $f_{i-1}: M \to N$ is a C^r map which is of class C^p on $W_1 \cup \ldots \cup W_{i-1}$ and approximates f within $\delta(1 - 1/2^{i-1})$. Consider $g_{i-1} = k_i f_{i-1} h_i^{-1}$, which carries $h_i(U_i)$ into the open subset $k_i(\mathcal{O}_i)$ of R^n or H^n. We apply Lemma 4.1 to obtain a C^r map g_i : $h_i(U_i) \to R^n$ or H^n, which equals g_{i-1} outside $h_i(V_i)$ and is of class C^∞ on $h_i(W_i)$. We need Exercise (a) of 4.1 if both M and N have boundaries. Let g_i be a close enough approximation to g_{i-1} that it carries $h_i(U_i)$ into $k_i(\mathcal{O}_i)$. Then f_i is well-defined by the equations

$$f_i(x) = f_{i-1}(x) \quad \text{for } x \text{ outside } \bar{V}_i$$
$$f_i(x) = k_i^{-1} g_i h_i(x) \quad \text{for } x \text{ in } U_i \quad .$$

Also, f_i is of class C^p on $W_1 \cup \ldots \cup W_i$. (Recalling that g_i is of class C^p on any open set where g_{i-1} is.) Furthermore, if we require $|g_i(y) - g_{i-1}(y)|$ and $|Dg_i(y) - Dg_{i-1}(y)|$ to be sufficiently small, we can make f_i a $\delta(x)/2^i$ approximation to f_{i-1}. (Here we use Lemma 3.6, of course.) Define $h(x) = \lim_{i \to \infty} f_i(x)$; h is well-defined because $f_i(y) = f_{i+1}(y) = \ldots$ on some neighborhood of x, for i sufficiently large. Also, h is of class C^p and is a δ-approximation to f.

To construct the differentiable homotopy between f and h, we preceed as follows: Lemma 4.1 gives us a C^r differentiable homotopy between g_{i-1} and g_i. From this we obtain a C^r differentiable homotopy $F_i(x, t)$ between $f_{i-1}(x)$ and $f_i(x)$; F_i will be a $\delta/2^i$ approximation to f_{i-1} for each fixed t; F_i is constant in t, for x outside V_i. Define

$$F(x, t) = F_1(x, t) \quad \text{for } t \leq 1$$
$$F(x, t) = F_{i+1}(x, t-i) \quad \text{for } i \leq t \leq i + 1 \quad .$$

Then F is of class C^r on $M \times R$; and for any compact subset B of M, there is a number n such that $F(x, t) = h(x)$ for $t > n$ and x in B. Define

$$f_t(x) = f(x) \quad \text{for } t \leq 0$$
$$f_t(x) = F(x, \tan(\pi t)) \quad \text{for } 0 \leq t < 1/2$$
$$f_t(x) = h(x) \quad \text{for } 1/2 \leq t \quad .$$

(a) Exercise. Strengthen the preceding theorem as follows: Let A be a closed subset of M and let f already be of class C^p in a neighborhood U of A. Then we may add the following to the conclusion of the theorem: (3) $f_1(x) = f(x)$ and $f_t(x) = f(x)$ for each x in A.

Hint: In the above proof, take $\{U_i\}$ to be a refinement of the covering $\{U, M - A\}$ of M, and choose $g_i = g_{i-1}$ if $U_i \subset U$.

(b) Exercise. Assume the hypotheses of the preceding theorem. Let B be a closed subset of the open set U of M. Prove there is a C^r map $h : M \to N$ such that h is of class C^p in a neighborhood of B and equals f outside U, and (1) and (2) of the preceding theorem hold.

(c) Exercise. Consider the preceding theorem in the case $r = 0$. Show that the results still hold, except that h and f_t are, of course, only C^0 δ-approximations to f.

4.3 Corollary. Let M and N be manifolds of class C^p; let $f : M \to N$ be a C^r immersion $(1 \leq r \leq p \leq \infty)$. There is a C^p immersion $f_1 : M \to N$, and a C^r differentiable homotopy f_t between f and f_1.

Proof. This follows from Theorems 3.10 and 4.2. We remind the reader that by our convention (3.9), f_t is required to be an immersion for each t.

4.4 Corollary. Let M and N be manifolds of class C^p; let $f : M \to N$ be a C^r imbedding $(1 \leq r \leq p \leq \infty)$. There is a C^p imbedding $f_1 : M \to N$, and a C^r differentiable isotopy between f and f_1.

4.5 Corollary. Let M and N be non-bounded manifolds of class C^p; let $f : M \to N$ be a C^r diffeomorphism $(1 \leq r \leq p \leq \infty)$. There is a C^p diffeomorphism $f_1 : M \to N$, and a C^r differentiable isotopy f_t between f and f_1; f_t is a diffeomorphism for each t.

(a) Exercise*. State and prove the stronger forms of these three corollaries, obtained by applying the results given in Exercises (a) and (b) of 4.2.

(b) Exercise*. Let M and N be manifolds of class C^r; let f_0 and f_1 be C^r maps of M into N; let f_t be a C^1 differentiable homotopy between f_0 and f_1. Let $\delta(x) > 0$. Prove there is a C^r

differentiable homotopy F_t between f_0 and f_1 such that F_t is a δ-approximation to f_t for each t.

4.6 Problem*. Let M and N be manifolds of class C^1; let f_0 and f_1 be C^1 maps of M into N; let f_t be a C^1 regular homotopy between them. Let $\delta(x) > 0$. Prove there is a C^1 differentiable homotopy F_t between them such that F_t is a δ-approximation to f_t, for each t.

Outline: Consider first the following problem: For each t, let f_t be a C^1 map of the open subset U of H^m or R^m into H^n or into R^n; suppose f_t is a C^1 regular homotopy between f_0 and f_1. Let A be a compact subset of U; let Ψ be as in 4.1; let $g_t(x) = \Psi(x) \cdot f_t(x)$. Define

$$h_t(x) = \int_{-\epsilon}^{+\epsilon} \varphi(s)\, g_{t+s}(x)\, ds \quad ,$$

for suitably small ϵ, where φ is positive on $(-\epsilon, \epsilon)$ and 0 outside, and $\int_{-\epsilon}^{\epsilon} \varphi = 1$. Let $F_t(x) = f_t(x) \cdot (1 - \Psi(x)) + h_t(x)$. Then F_t is a C^1 differentiable homotopy between f_0 and f_1 on some neighborhood of A, and equals f_t outside some neighborhood of A.

4.7 Lemma. Let U, V, and W be bounded open subsets of R^m, with $\bar{W} \subset V$ and $\bar{V} \subset U$. Let π be the projection of R^n onto R^m. Suppose $f : U \to R^n$ is a C^r map such that $\pi f : U \to R^m$ is an imbedding. Given $\delta(x) > 0$, there is a C^r imbedding $h : U \to R^n$ such that

(1) h equals f outside V.

(2) h is a δ-approximation to f.

(3) $h(W)$ is a C^∞ submanifold of R^n.

(4) If U_1 is an open subset of U and $f(U_1)$ is a C^∞ submanifold of R^n, then $h(U_1)$ is also a C^∞ submanifold of R^n.

Proof. The restriction of π to $f(U)$ is a homeomorphism of $f(U)$ onto an open subset $\mathcal{O}$ of R^m; let $g : \mathcal{O} \to R^n$ be the inverse of this map. Then

$$g(x^1, \ldots, x^m) = (x^1, \ldots, x^m, g^{m+1}(x), \ldots, g^n(x)) \quad ,$$

and g is of class C^r, since it equals $f(\pi f)^{-1}$.

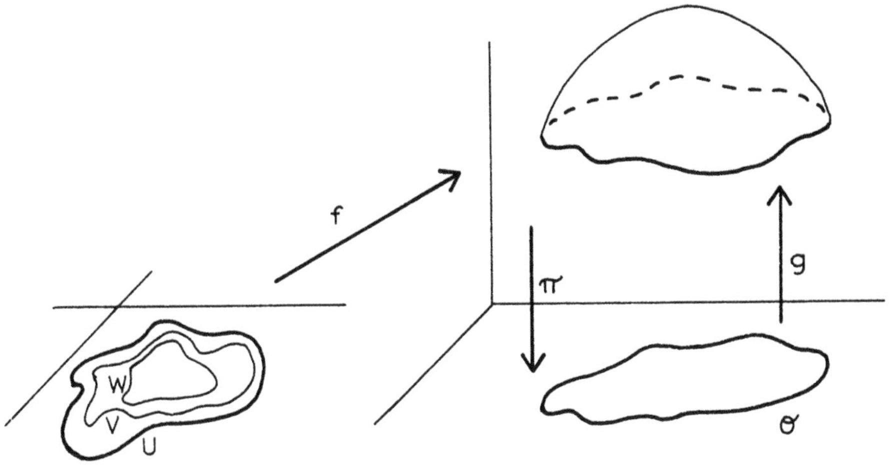

Further, if $f(U_1)$ is a C^∞ submanifold of R^n, the restriction of g to $\pi f(U_1)$ is of class C^∞. (The converse is easy (2.2).) For impose on $f(U_1)$ its induced C^∞ structure, and note that $\pi|f(U_1)$ is a C^∞ map relative to this structure. The map g is the inverse of this, which is C^∞, followed by the inclusion of $f(U_1)$ into R^n, which is C^∞.

Consider $g_0 : \mathcal{O} \to R^{n-m}$, defined by the equation

$$g_0(x) = (g^{m+1}(x), \ldots, g^n(x)) \quad .$$

Using 4.1, choose $\tilde{g}_0(x)$ to be an ϵ-approximation to g_0 which is of class C^∞ on $\pi f(W)$ and equals g_0 outside $\pi f(V)$. We further require that $\tilde{g}_0$ be of class C^∞ on any open subset of $\mathcal{O}$ where g_0 is C^∞. Let $\tilde{g} : \mathcal{O} \to R^n$ be defined by the equation $\tilde{g}(x) = (x, \tilde{g}_0(x))$; it is an ϵ-approximation to g.

Define $h : U \to R^n$ by the equation $h(x) = \tilde{g}\pi f(x)$. By 3.7, if ϵ is properly chosen, h will be a δ-approximation to $f = g\pi f$.

<u>4.8 Theorem</u>. Let M be a non-bounded C^r manifold $(1 \le r)$. Let $f : M \to R^n$ be a C^r imbedding. Let $\delta(x) > 0$. There is a C^r imbedding $h : M \to R^n$ which is a δ-approximation to f, such that $h(M)$ is a C^∞ submanifold of R^n.

<u>Proof</u>. Let δ be small enough that any δ-approximation to f is an imbedding. For each x in M, $df(x)$ has rank m. Then for some projection π of R^n onto a coordinate m-plane, the map $d(\pi f)(x)$ has rank m,

so that πf is a C^r diffeomorphism of a neighborhood of x onto an open set in the coordinate m-plane, by the inverse function theorem. Choose a covering of M by sets C_i such that for some coordinate system (U_i, h_i) about C_i, $h_i(C_i)$ equals the unit m-ball, such that the sets Int C_i cover M, and such that for some projection π_i of R^n onto a coordinate m-plane, $\pi_i f|C_i$ is an imbedding. Let δ_i be a number such that any δ_i-approximation to $\pi_i f|C_i$ is an imbedding; let $\delta(x)$ be small enough that $\delta(x) < \delta_i$ for x in C_i. Then if $f': M \to R^n$ is a δ-approximation to f, then $\pi_i f'$ is a δ-approximation to $\pi_i f$, so that $\pi_i f'|C_i$ is an imbedding. Let W_i be an open covering of M, with $\bar{W}_i$ contained in the open set V_i, and $\bar{V}_i \subset \text{Int } C_i$.

Let $f_0 = f$. As an induction hypothesis, suppose $f_{j-1}: M \to R^n$ is a C^r map which is a $(1 - 1/2^{j-1})\delta(x)$ approximation to f such that $f_{j-1}(W_k)$ is a C^∞ submanifold of R^n, for $k < j$. We then construct a map f_j. Apply the preceding lemma to the map

$$f_{j-1}h_j^{-1} : h_j(\text{Int } C_j) \to R^n \quad ,$$

to obtain a map $f_j: M \to R^n$ which is a $\delta/2^j$ approximation to f_{j-1}, which equals f_{j-1} outside V_j, and such that $f_j(W_j)$ is a C^∞ submanifold of R^n. Condition (4) of the preceding lemma guarantees we can choose f_j so that $f_j(W_k)$ is a C^∞ submanifold of R^n for $k < j$.

As before, we let $h(x) = \lim_{j \to \infty} f_j(x)$, and note that h satisfies the conditions of the theorem.

<u>4.9 Corollary</u>. If $\mathfrak{D}$ is a C^r differentiable structure on the non-bounded manifold M, $\mathfrak{D}$ contains a C^∞ structure.

(a) Exercise. Let M be a non-bounded C^r manifold; let N be a non-bounded C^∞ manifold. Let $\delta(x) > 0$. If $f : M \to N$ is a C^r imbedding, prove there is a C^r imbedding $h : M \to N$ which is a δ-approximation to f, such that $h(M)$ is a C^∞ submanifold of N, and h is C^r differentiably isotopic to f.

§5. Manifolds with Boundary.

There are additional technicalities involved in proving our two main theorems 4.5 and 4.9, for manifolds with boundary.

First, we need to prove the local retraction theorem (5.5), which states that a non-bounded C^r submanifold M of euclidean space has a neighborhood which is retractable onto M by a C^r retraction. If $r > 1$, it is easy to find such a retraction of class C^{r-1}; roughly, one takes the plane normal to M at x and collapses it onto x. To construct a retraction of class C^r requires more work; one needs first to study the Grassmann manifolds (5.1 - 5.4). One also needs a topological lemma (5.7).

From this, we can prove the product neighborhood theorem, which states that $\text{Bd } M$ has a neighborhood in M which is diffeomorphic with $\text{Bd } M \times [0, 1)$. This in turn is used to construct a differentiable structure on $D(M)$, the double of M (5.8 - 5.10).

The theorems for M then reduce to the corresponding theorems for the non-bounded manifold $D(M)$ (5.11 - 5.13).

5.1 Definition. The Grassmann manifold $G_{p,n}$ is the set of all n-dimensional subspaces of R^{n+p}.

Let $M(p,q)$ denote the set of all $p \times q$ matrices; $M(p,q;r)$ denotes those having rank r. The rows of any matrix A of $M(n,n+p;n)$ are a set of n independent vectors of R^{n+p}, so they determine an element of $G_{p,n}$ which we denote by $\lambda(A)$. Further, two matrices A and B determine the same element of $G_{p,n}$ if and only if the rows of each are linear combinations of the rows of the other; i.e., if $A = CB$ for some non-singular $n \times n$ matrix C.

$M(n,n+p;n)$ is an open subset of $R^{n(n+p)}$ (see 3.10), so that it has a natural topology and C^∞ differentiable structure. Let $G_{p,n}$ be given the identification topology: V is open in $G_{p,n}$ if and only if $\lambda^{-1}(V)$ is open in $M(n, n+p; n)$.

We note that $\lambda: M(n, n+p; n) \to G_{p,n}$ is an open map. For let U be open in $M(n, n+p; n)$. For any C in $M(n, n; n)$, the map $C : A \to CA$

is a homeomorphism of $M(n, n+p; n)$ with itself, so that $C(U)$ is open in $M(n, n+p; n)$. $\lambda^{-1}(\lambda(U))$ is the union of the sets $C(U)$ for all C in $M(n, n; n)$, so $\lambda(U)$ is open.

5.2 Theorem. $G_{p,n}$ is a non-bounded C^∞ manifold of dim pn; the map $\lambda : M(n,n+p;n) \to G_{p,n}$ is of class C^∞.

Proof. (1) $G_{p,n}$ is locally euclidean. Let us explain first the geometric idea of the proof. If $\mathcal{P}$ is an n-plane through the origin in R^{n+p}, there is some projection π of R^{n+p} onto a coordinate n-plane which is a linear isomorphism on $\mathcal{P}$. Let U be the set of all planes on which π is a linear isomorphism. Each plane of U uniquely determines an n-tuple of vectors $(\vec{v}_1, \ldots, \vec{v}_n)$ which lie in it and project under π into the natural unit basis vectors for the coordinate n-plane. Conversely, each such n-tuple determines a plane of U. Thus, each vector $\vec{v}_i$ has p components which may be chosen arbitrarily; since there are n such vectors, U is homeomorphic with R^{pn}.

More precisely, let $\lambda(A)$ be the general element of $G_{p,n}$. Some $n \times n$ submatrix of A has rank n; suppose it consists of the first n columns. Let U be the set of matrices of the form $[P\ Q]$ where P is $n \times n$ and non-singular. Then U is open in $R^{n(n+p)}$ and hence in $M(n,n+p;n)$. Hence $V = \lambda(U)$ is open in $G_{p,n}$.

Let Ψ map the arbitrary matrix Q of $M(n,p)$ into the matrix $[I\ Q]$ of $M(n,n+p)$; let $\Psi_0 = \lambda\Psi$. Then Ψ_0 is a continuous map of $M(n,p)$ into $G_{p,n}$. It is 1-1, since $\lambda[I\ Q_1] = \lambda[I\ Q_2]$ only if $[I\ Q_1] = C[I\ Q_2]$; it maps $M(n,p)$ onto V, since $\lambda[P\ Q] = \lambda[I\ \ P^{-1}Q] = \Psi_0(P^{-1}\ Q)$.

Let $\varphi : U \to M(n,p)$ be defined by the equation $\varphi([P\ Q]) = P^{-1}Q$. Now φ is constant on each set $\lambda^{-1}(\mathcal{P})$, since the equation $\lambda([P_1\ Q_1]) = \lambda([P_2\ Q_2])$ means that $[P_1\ Q_1] = C[P_2\ Q_2]$, whence $P_1^{-1}Q_1 = (CP_2)^{-1}(CQ_2) = P_2^{-1}Q_2$. Hence φ induces a continuous map φ_0 of V into $M(n,p)$; φ_0 is the inverse of Ψ_0, since $\varphi_0\Psi_0(Q) = \varphi([I\ Q]) = I^{-1}Q = Q$. Hence Ψ_0 is a homeomorphism.

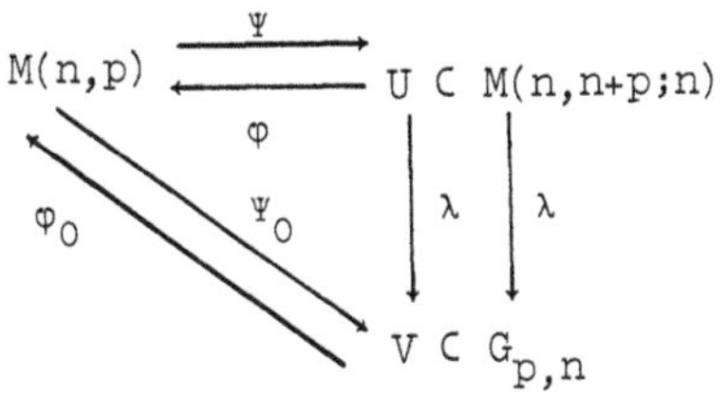

(2) $G_{p,n}$ has a C^∞ differentiable structure. We need only prove that two coordinate neighborhoods (V, φ_0) of the type just constructed have class C^∞ overlap. Let $\tilde{\varphi}, \tilde{\Psi}, \tilde{U}, \tilde{V}$ be the corresponding maps and open sets which determine another coordinate neighborhood. Then $\tilde{\Psi}$ takes the columns of the matrix Q and distributes them in a certain way among the columns of the identity matrix I, so that $\tilde{\Psi}$ is certainly of class C^∞. Likewise φ is of class C^∞ on the open set U of $M(n,n+p;n)$, since $\varphi([P\ Q]) = P^{-1}Q$. Hence $\varphi_0\tilde{\Psi}_0 = \varphi\tilde{\Psi}$ is of class C^∞ (on the open set $\tilde{\varphi}(\tilde{U} \cap U)$ where it is defined).

(3) $G_{p,n}$ is Hausdorff. For let $\lambda(A)$ and $\lambda(B)$ be distinct subspaces of $G_{p,n}$. Then the matrix $\left[\begin{smallmatrix} A \\ B \end{smallmatrix} \right]$ must have rank at least $n+1$. Choose $\epsilon > 0$ so that any $2n \times (n+p)$ matrix within ϵ of this one has rank at least $n+1$. Let U denote the ϵ-neighborhood of A, and V, the ϵ-neighborhood of B, in $M(n, n+p; n)$. Then $\lambda(U)$ and $\lambda(V)$ are disjoint neighborhoods of $\lambda(A)$ and $\lambda(B)$, respectively, since λ is an open map.

(4) $G_{p,n}$ has a countable basis. Indeed, $G_{p,n}$ is covered by $(n+p)!/n!p!$ coordinate systems of the type constructed above.

(a) Exercise*. Show that $G_{p,n}$ is compact.

(b) Exercise*. Show that there is a C^∞ diffeomorphism of $G_{p,n}$ onto $G_{n,p}$.

<u>5.3 Lemma</u>. Let $f : M \to G_{p,n}$ be a C^r map. Given x in M, there is a neighborhood U of x and a C^r map $f_*: U \to M(n,n+p;n)$ such that $\lambda f_* = f$. (f_* is called a <u>lifting of</u> f <u>over</u> U.)

<u>Proof</u>. Let (V, φ_0) be a coordinate system on $G_{p,n}$, as in 5.2, such that $f(x)$ lies in V. Now $\varphi_0: V \to M(n,p)$ is a C^∞ diffeomorphism.

Simply define

$$f_*(x) = \Psi(\varphi_0(f(x))) = [I \ \varphi_0 f(x)] \ .$$

(a) Exercise*. Let (V, φ_0) be as in 5.2. Show that there is a diffeomorphism g of $\lambda^{-1}(V)$ with $V \times M(n, n; n)$ such that λg^{-1} is the natural projection of $V \times M(n, n; n)$ onto V.

Let $(\tilde{V}, \tilde{\varphi}_0)$ be another coordinate system on $G_{p,n}$, and $\tilde{g}$ the corresponding diffeomorphism. For each x in $V \cap \tilde{V}$, let a map h_x, carrying $M(n,n;n)$ into itself, be defined by the equation

$$(x, h_x(P)) = \tilde{g}\, g^{-1}(x, P) \ .$$

Show that h_x is merely multiplication by a non-singular matrix A_x, and show that the map $x \to A_x$ is continuous. This will show that $\lambda : M(n,n+p;n) \to G_{p,n}$ is a <u>principal fibre bundle</u> with fibre and structural group $M(n,n;n)$. (See [12].)

<u>5.4 Definition</u>. Let $f : M \to R^n$ be a C^r immersion. If $\vec{v}$ is tangent to M at x, then $df(\vec{v})$ is a tangent vector to R^n at $f(x)$. If we identify $T(R^n)$ with $R^n \times R^n$, then $df(\vec{v}) = (f(x), d_0 f(\vec{v}))$, as in 3.5. As $\vec{v}$ ranges over the tangent space at x, $d_0 f(\vec{v})$ ranges over an m-plane through the origin in R^n. The <u>tangent map</u> t of the immersion f will be the map $t : M \to G_{p,m}$ which assigns this m-plane to the point x. ($n = m + p$, of course.)

Let (U, h) be a coordinate system on M. Then locally the map t is given by the equation $t(x) = \lambda(D(fh^{-1}))$, where the matrix $D(fh^{-1})$ is evaluated at $h(x)$, and λ is the projection of $M(m,m+p;m)$ onto $G_{p,m}$. Hence t is of class C^{r-1}.

Similarly, one has the <u>normal map</u> of the immersion, $n : M \to G_{m,p}$, which carries x into the orthogonal complement of $t(x)$. Now the matrix $D(fh^{-1})$ is non-singular; suppose it is of the form $[P(x)\ Q(x)]$ where P is $m \times m$ and non-singular. Then locally n is given by the equation $n(x) = \lambda([R(x)\ I])$, where $R = -(P^{-1}Q)^{tr}$. Hence n is also of class C^{r-1}.

5.5 Theorem (local retraction theorem). Let $f : M \to R^n$ be a C^r imbedding; M non-bounded. There is a neighborhood W of $f(M)$ and a C^r retraction $r : W \to f(M)$. (We recall this means that $r(y) = y$ for y in $f(M)$.)

Proof. Consider the normal map $n : M \to G_{m,p}$ and the tangent map $t : M \to G_{p,m}$ of the imbedding f $(p = n - m)$. Both are of class C^{r-1}. There is a fine C^0 neighborhood of n such that for any map $\tilde{n}$ which lies in this neighborhood, $\tilde{n}(x)$ and $t(x)$ are independent subspaces of R^n, i.e., their intersection is the zero vector (see Exercise (a)). Choose a map $\tilde{n} : M \to G_{p,m}$ of class C^r which lies in this neighborhood, using Theorem 4.2. (In the case $r = 1$, n is only of class C^0, and we need to use Exercise (c) of that section.) Let $\tilde{t}(x)$ be the orthogonal complement of $\tilde{n}(x)$; $\tilde{t}$ is also a map of class C^r.

Let E be the subset of $M \times R^n$ consisting of pairs $(x, \vec{v})$ such that $\vec{v}$ lies in the subspace $\tilde{n}(x)$. (If we used the map n instead of $\tilde{n}$, E would be what is called the normal bundle of the imbedding.) We prove that E is a C^r submanifold of $M \times R^n$ of dimension n.

Given x in M; let $\tilde{n}_*$ and $\tilde{t}_*$ be C^r liftings of $\tilde{n}$ and $\tilde{t}$ in a neighborhood of x. Then $\tilde{n}_*(x) = A(x)$ is a $p \times n$ matrix, and $\tilde{t}_*(x) = B(x)$ is an $m \times n$ matrix; the rows of A and B jointly form a linearly independent set. Let (U, h) be a coordinate system about x such that $\tilde{n}_*$ and $\tilde{t}_*$ are defined on U; let $h(x) = (u^1, \ldots, u^m)$. Define $g : U \times R^n \to R^m \times R^n$ by the equation

$$g(x,(v^1,\ldots,v^n)) = \left(h(x),\ [v^1 \ldots v^n] \cdot \begin{bmatrix} A(x) \\ B(x) \end{bmatrix}^{-1}\right) .$$

Then $(U \times R^n, g)$ is a C^r coordinate system on $M \times R^n$. Now if $(x, \vec{v})$ lies in E, then $\vec{v}$ is some linear combination of the rows of $A(x)$, so that

$$\vec{v} = [y^1 \ldots y^p\ 0 \ldots 0] \cdot \begin{bmatrix} A(x) \\ B(x) \end{bmatrix}$$

for some choice of $y^1, \ldots, y^p$. Conversely, if $\vec{v}$ is of this form, then $(x, \vec{v})$ lies in E. As a result, $(x, \vec{v})$ lies in E if and only if $g(x, \vec{v})$ lies in $h(U) \times R^p$. Hence the coordinate map g carries the intersection of E with $U \times R^n$ onto the open subset $h(U) \times R^p$ of R^{m+p}, so that E is a C^r submanifold of $M \times R^n$.

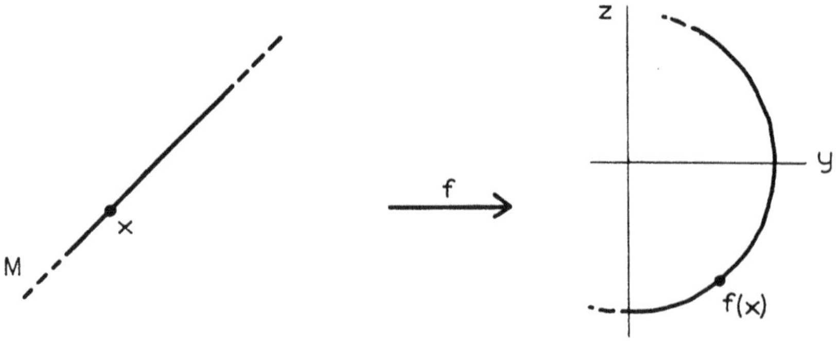

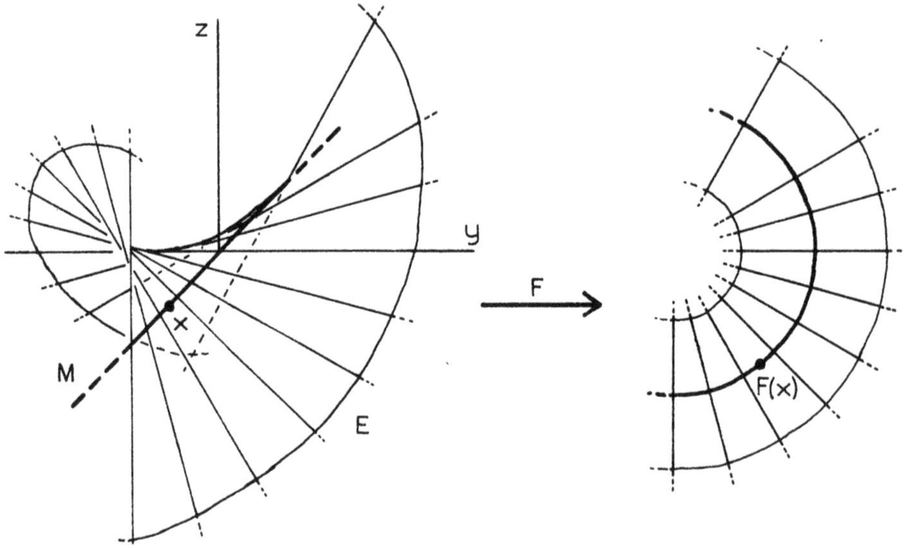

Consider the C^r map of $M \times R^n \to R^n$ which carries $(x, \vec{v})$ into $f(x) + \vec{v}$; let F be the restriction of this map to E.

We prove that dF has rank n at each point of $M \times 0$. Use the coordinate system $g(x, \vec{v}) = (u^1, \ldots, u^m, y^1, \ldots, y^p, 0, \ldots, 0)$ on E which was constructed above. Then

$$Fg^{-1}(u, y) = fh^{-1}(u) + [y^1 \ldots y^p] \cdot A(h^{-1}(u)) \quad ,$$

so that

$$\partial(Fg^{-1})/\partial y = A(h^{-1}(u))^{tr} \; ,$$

and

$$\partial(Fg^{-1})/\partial u^k = \partial(fh^{-1})/\partial u^k + ([y^1 \ldots y^p] \cdot \partial(Ah^{-1})/\partial u^k)^{tr} \; .$$

The last term vanishes at points of $g(M \times 0)$, since $y = 0$ at such points. Hence

$$D(Fg^{-1}) = [D(fh^{-1})(u) \quad A(h^{-1}(u))^{tr}]$$

at points of $g(M \times 0)$. The columns of $D(fh^{-1})$ span the tangent plane $t(x)$;

the rows of A span the plane $\tilde{n}(x)$; by choice of $\tilde{n}$ these spaces are independent.

Now, $F : E \to R^n$ is a homeomorphism when restricted to the subspace $M \times 0$; furthermore, each point of $M \times 0$ has a neighborhood which F maps homeomorphically onto an open subset of R^n (by the inverse function theorem). It follows that there is a neighborhood of $M \times 0$ in E which F maps homeomorphically onto an open set in R^n (see Lemma 5.7). There is also a neighborhood of $M \times 0$ in E on which dF has rank n; let V be the intersection of these neighborhoods and set $W = F(V)$.

There is a natural projection π of $M \times R^n$ onto M. Since F is a C^r diffeomorphism of V onto W, $F \pi F^{-1} = r$ is the required retraction of W onto $f(M)$.

(a) Exercise. Construct the fine C^0 neighborhood of the normal map n required in the preceding proof.

Hint: Use the fact that λ is an open map to construct neighborhoods of $n(x_0)$ and $t(x_0)$ consisting of subspaces which are independent.

5.6 Corollary. Let $f : M \to N$ be a C^r imbedding; M non-bounded. There is a neighborhood W of $f(M)$ in N and a C^r retraction $r : W \to f(M)$.

Proof. Let $g : N \to R^q$ be a C^r imbedding. There is a neighborhood U of $gf(M)$ in R^q and a C^r retraction r_0 of U onto $gf(M)$. Then $g^{-1} r_0 g$ is the required retraction, defined on $W = g^{-1}(U)$.

(a) Exercise.* Let $f : M \to N$ be a C^r immersion, with $r \geq 2$; M and N non-bounded. The normal bundle E of f may be described most simply when N is a submanifold of R^p having the derived Riemannian metric. In this case, let E be the subspace of $M \times R^p$ consisting of pairs $(x, \vec{v})$ such that $\vec{v}$ is a p-tuple representing a vector tangent to N and normal to $df(\vec{w})$ for each $\vec{w}$ tangent to M at x. (In the more general case, E is described as a subset of $M \times T(N)$.) Then E is a C^{r-1} submanifold of $M \times R^p$. Prove the tubular neighborhood theorem:

If $f : M \to N \subset R^p$ is a C^r imbedding, there is a neighborhood U of $M \times 0$ in the normal bundle E and a C^{r-1} diffeomorphism h of U with a neighborhood of $f(M)$ in N such that $h|M \times 0 = f$.

Hint: Let $F : E \to R^p$ be the map $F(x, \vec{v}) = f(x) + \vec{v}$; let r be a retraction of a neighborhood of N onto N; let $h = rF$.

5.7 Lemma. Let A be a closed subset of the locally compact space X. Let $f: X \to Y$ be a homeomorphism when restricted to A; suppose each point x of A has a neighborhood U_x which f maps homeomorphically onto an open subset of Y. Then f is a homeomorphism of some neighborhood of A onto an open subset of Y. (X and Y are separable metric, as always.)

Proof. Let U be the union of the neighborhoods U_x. Then f is locally 1-1 at each point of U. Furthermore, if B is any subset of U such that $f|B$ is 1-1, then $f|B$ is a homeomorphism. For let $f(x_n) \to f(x)$, where x_n and x are in B. Now $f(U_x)$ is open in Y, so it contains all elements of the sequence $f(x_n)$ from some n onwards. Since $f|U_x$ is a homeomorphism, x_n must converge to x.

(1) If C is a compact subset of U and f is 1-1 on C, we prove there is a neighborhood of C on whose closure f is 1-1. Let U_n be the ϵ/n-neighborhood of C, where ϵ is small enough that $\bar{U}_1$ is compact and lies in U. If f is 1-1 on no set $\bar{U}_n$, there are points x_n and y_n of $\bar{U}_n$ such that $f(x_n) = f(y_n)$. By passing to subsequences and renumbering, we may assume $x_n \to x$ and $y_n \to y$; x and y will be points of C. Then $f(x) = f(y)$; since f is 1-1 on C, $x = y$. But f is locally 1-1 at x, so we could not have $f(x_n) = f(y_n)$ for large n.

(2) If C is a compact subset of U and f is 1-1 on $C \cup A$, we prove there is a neighborhood V of C such that f is 1-1 on $\bar{V} \cup A$. Let U_n be the ϵ/n-neighborhood of C, where ϵ is small enough that $\bar{U}_1$ is compact and lies in U, and so that f is 1-1 on $\bar{U}_1$ (using (1)). If f is 1-1 on no set $\bar{U}_n \cup A$, there are points x_n of $\bar{U}_n - A$ and y_n of $A - \bar{U}_n$ such that $f(x_n) = f(y_n)$. By passing to subsequences and renumbering, we may assume $x_n \to x$, where x is a point of C. Then $f(y_n) \to$

$f(x)$. Now $f|C \cup A$ is a homeomorphism, so that $y_n \to x$. But f is locally 1-1 at x, so we could not have $f(x_n) = f(y_n)$ for large n.

(3) We now prove the lemma. Let A be the union of the increasing sequence $A_0 \subset A_1 \subset \ldots$ of compact sets, where A_0 is empty. Define V_0 to be the empty set. In general, suppose V_n is a neighborhood of A_n, such that $\bar{V}_n$ is a compact subset of U and f is 1-1 on $\bar{V}_n \cup A$. By (2), we may choose a neighborhood V_{n+1} of the compact set $\bar{V}_n \cup A_{n+1}$ such that $\bar{V}_{n+1}$ is a compact subset of U and f is 1-1 on $\bar{V}_{n+1} \cup A$.

Let V be the union of the sets V_n. Then f is 1-1 on the neighborhood V of A.

5.8 Lemma. Let M be a C^r manifold. There is a non-negative real-valued C^r function g on M such that for each x in $\mathrm{Bd}\, M$, $g(x) = 0$ and $dg(x)$ has rank 1.

Proof. Let $\{(U_i, h_i)\}$ be a locally-finite covering of M by coordinate systems; let $\{\varphi_i\}$ be a partition of unity dominated by the covering $\{U_i\}$. Let $\dim M = m$. For each i, the mth coordinate function $h_i^m(x)$ is zero for x in $U_i \cap \mathrm{Bd}\, M$. Furthermore, if $k(x) = (u^1, \ldots, u^m)$ is another coordinate sustem about the point x, then $\partial(h_i^m k^{-1})/\partial u^m$ is positive at $k(x)$: For $h_i k^{-1}$ is a non-singular transformation of an open set in H^m about $k(x)$ onto an open set in H^m, and $h_i^m k^{-1}(u^1, \ldots, u^{m-1}, 0) \equiv 0$, so that $\partial(h_i^m k^{-1})/\partial u^j = 0$ at points of R^{m-1}, for $j < m$. Since $D(h_i k^{-1})$ is non-singular, $\partial(h_i^m k^{-1})/\partial u^m$ must be non-zero at points of R^{m-1}; since $h_i^m(x)$ is non-negative for all x, this derivative must be positive at points of R^{m-1}.

Let $g(x) = \Sigma_i \varphi_i(x) \cdot h_i^m(x)$. Then $g(x) = 0$ whenever x is in $\mathrm{Bd}\, M$. Furthermore, if $k(x) = (u^1, \ldots, u^m)$ is a coordinate system on M, then

$$\partial(gk^{-1})/\partial u^m = \Sigma_i\, h_i^m \cdot \partial(\varphi_i k^{-1})/\partial u^m \; + \; \Sigma_i\, \varphi_i \cdot \partial(h_i^m k^{-1})/\partial u^m \quad .$$

Whenever x is in $\mathrm{Bd}\, M$, the first term vanishes, because $h_i^m(x) = 0$; the second term is strictly positive. Hence $dg(x)$ has rank 1 if x is in $\mathrm{Bd}\, M$.

5.9 Theorem. Let M be a C^r manifold. There is a C^r diffeomorphism p of a neighborhood of $\operatorname{Bd} M$ with $\operatorname{Bd} M \times [0, 1)$ such that $p(x) = (x, 0)$ whenever x is in $\operatorname{Bd} M$. Such a neighborhood is called a product neighborhood of the boundary.

Proof. By 5.6, there is a neighborhood W of $\operatorname{Bd} M$ in M and a C^r retraction $r : W \to \operatorname{Bd} M$. By 5.8, there is a non-negative C^r function g on M such that if x is in $\operatorname{Bd} M$, $g(x) = 0$ and $dg(x)$ has rank 1. Define $f : W \to \operatorname{Bd} M \times [0, \infty)$ by the equation $f(y) = (r(y), g(y))$. If x is in $\operatorname{Bd} M$, then $f(x) = (x, 0)$ and $df(x)$ is non-singular: For let $k(x) = (u^1, \ldots, u^m)$ be a coordinate system about x; let $h = (k|\operatorname{Bd} M) \times i$.

$$D(hfk^{-1}) = \begin{bmatrix} I & * \\ 0 & \partial(gk^{-1})/\partial u^m \end{bmatrix} .$$

One uses Theorem 1.11 to prove that each x in $\operatorname{Bd} M$ has a neighborhood U_x which f carries homeomorphically onto an open subset of $\operatorname{Bd} M \times [0, \infty)$. Since f is a homeomorphism when restricted to $\operatorname{Bd} M$, it follows from 5.7 that f is a homeomorphism of some neighborhood U of $\operatorname{Bd} M$ in W onto a neighborhood V of $\operatorname{Bd} M \times 0$ in $\operatorname{Bd} M \times [0, \infty)$. Using Exercise (a) of 2.6, we may construct a C^r function $\delta(x) > 0$ on $\operatorname{Bd} M$ such that the point (x, t) of $\operatorname{Bd} M \times [0, \infty)$ lies in V if $t < \delta(x)$. Let s be the diffeomorphism of $\operatorname{Bd} M \times [0, \infty)$ onto itself which carries (x, t) into $(x, t/\delta(x))$, and let $p = sf$. Then $p(U)$ contains $\operatorname{Bd} M \times [0, 1)$, and a restriction of p will satisfy the demands of the theorem.

5.10 Definition. Let M be a C^r manifold with non-empty boundary. Recall that the double of M, $D(M)$, is the union of $M_0 = M \times 0$ and $M_1 = M \times 1$, with $(x, 0)$ and $(x, 1)$ identified whenever x is in $\operatorname{Bd} M$. We may impose a differentiable structure on $D(M)$ as follows: Let $p_0 \colon U_0 \to \operatorname{Bd} M_0 \times [0, 1)$ and $p_1 \colon U_1 \to \operatorname{Bd} M_1 \times (-1, 0]$ be product neighborhoods of the boundary in M_0 and M_1. Let U be the union of U_0 and U_1 in $D(M)$, and let $P : U \to \operatorname{Bd} M \times (-1, 1)$ be the homeomorphism induced by p_0 and p_1. A C^r differentiable structure on $D(M)$ is well-defined if we require (1) P to be a C^r diffeomorphism, and (2) the inclusions of M_0 and M_1 in $D(M)$ to be C^r imbeddings.

Now the differentiable structure on $D(M)$ depends strongly on the choice of the product neighborhoods p_0 and p_1. However, the differentiable manifolds arising from two such choices are diffeomorphic, as we shall prove later (6.3).

(a) Exercise*. Impose two distinct differentiable structures on $D(H^2)$ by using different choices for the product neighborhood of Bd H^2. Show that the resulting differentiable manifolds are diffeomorphic, but that the diffeomorphism may not be chosen as an arbitrarily good C^1 approximation to the identity on each copy of H^2.

(b) Exercise*. If M and N are C^r manifolds, show that $M \times N$ has a C^r differentiable structure such that each inclusion $M \to M \times y$ and $N \to x \times N$ is a C^r imbedding.

5.11 Theorem. If $\mathfrak{D}$ is a C^r differentiable structure on the manifold M, then $\mathfrak{D}$ contains a C^∞ structure.

Proof. We have proved the theorem in the case where Bd M is empty. Otherwise, let us consider the non-bounded manifold $D(M)$, and provide it with a C^r differentiable structure, as in 5.10. Choose a C^∞ structure contained in this, and let D' denote the corresponding differentiable manifold.

Consider the map P^{-1}: Bd $M \times (-1, 1) \to D(M)$ where P is the map defined in 5.10. The C^r structure on Bd M contains a C^∞ one; let us denote the resulting differentiable manifold by (Bd M)'. Then P^{-1} is a C^r diffeomorphism of the C^∞ manifold (Bd M)' $\times (-1, 1)$ with an open subset U of the C^∞ manifold D'.

Now there is a C^r diffeomorphism h of (Bd M)' $\times (-1, 1)$ with U, which is C^∞ in a neighborhood of (Bd M)' $\times 0$, and equals P^{-1} outside (Bd M)' $\times (-1/2, 1/2)$. (Here we use Exercise (b) of 4.2.) Then the set h(Bd $M \times 0$) is a C^∞ submanifold of D'.

Now h induces a C^r diffeomorphism f of D' onto itself; we merely define $f(x) = hP(x)$ if x is in U, and $f(x) = x$ otherwise. If $i : M \to M_0 \subset D'$ is the inclusion mapping, then fi is a C^r imbedding of M in D', and the set $fi(M)$ is a C^∞ submanifold of D'. Our result follows

5.12 Lemma. Let M and N be non-bounded C^p manifolds; let $f : M \to N$ be a C^r diffeomorphism. Identify N with $N \times 0$ in $N \times R$ for convenience. Given a positive function ϵ on $N \times R$, there is a positive function δ on M such that the following holds:

Let g be a C^p imbedding of M into $N \times R$ which is a δ-approximation to f. Then there is a C^p diffeomorphism h of $N \times R$ onto itself, such that

(1) h carries $g(M)$ into N.

(2) h is an ϵ-approximation to the identity, and equals the identity outside $N \times (-2/3, 2/3)$.

Proof. Let $\beta(t)$ be a monotonic C^∞ function which equals 1 for $t \leq 1/3$ and equals 0 for $t \geq 2/3$. Let $k \geq |\beta'(t)|$ for all t; then $k \geq 3$. Let π be the projection of $N \times R$ onto $N \times 0$. Let δ be small enough that $g(M)$ lies in $N \times (-1/3, 1/3)$, and small enough that πg is a diffeomorphism of M onto $N \times 0$. (By 3.7, πg may be made to approximate $\pi f = f$ as closely as desired by taking δ small enough.)

Then the general point of $g(M)$ is of the form $(y, \varphi(y))$, where y is in N and φ is a real function on N, with $-1/3 < \varphi(y) < 1/3$. Since $(y, \varphi(y)) = g((\pi g)^{-1}(y, 0))$, $\varphi(y)$ is a C^p function of y.

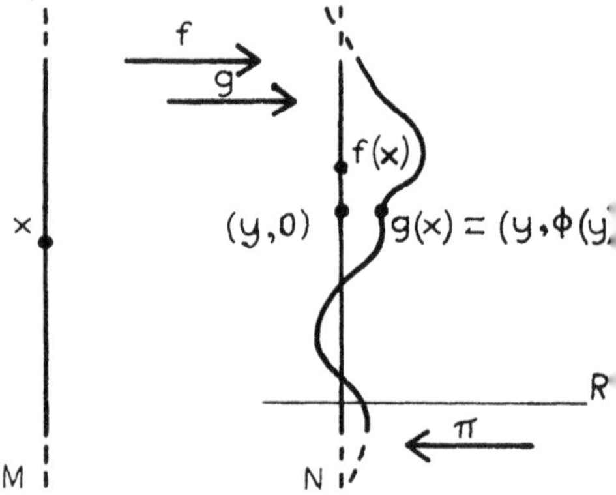

Define $h : N \times R \to N \times R$ by the equation

$$h(y, s) = (y, \Psi(y, s)),$$

where

$$\Psi(y, s) = s - \varphi(y)\beta(|s|).$$

Now $|\varphi(y)| < 1/3$, so when $s = \varphi(y)$, $\beta(|s|) = 1$. Hence h carries $g(M)$ into $N \times 0$. If $|s| > 2/3$, $\beta(|s|) = 0$ and h is the identity.

By proper choice of δ, we may make $\varphi(y)$ as close an approximation to the zero-function as desired. For this, we note that $(y, \varphi(y)) = g(\pi g)^{-1}(y, 0)$; using 3.7, we see that if g approximates f closely enough, πg approximates $\pi f = f$, $(\pi g)^{-1}$ approximates f^{-1}, and $g(\pi g)^{-1}$ approximates ff^{-1} = identity. Let δ be small enough that $\varphi(y)$ is an $\varepsilon_0(y)/2k$ approximation to the zero function, where $\varepsilon_0(y) = \min \varepsilon(y, s)$ for $-1 \leq s \leq 1$. N must be imbedded in some R^p before this makes sense; then $N \times R$ is imbedded in R^{p+1}. Now $|\varphi(y)| < \varepsilon_0(y)$, so

$$\| h(y, s) - (y, s) \| = |\Psi(y, s) - s| < \epsilon_0(y) \quad ,$$

as desired. Let $\vec{v}$ be a unit tangent to $N \times R$ at (y, s); let $\vec{v}_1$ and $\vec{v}_2$ be its components tangent to $N \times s$ and $y \times R$, respectively.

$$\| dh(\vec{v}) - \vec{v} \| \leq \| dh(\vec{v}_1) - \vec{v}_1 \| + \| dh(\vec{v}_2) - \vec{v}_2 \| \quad .$$

The second term is less than $\varepsilon_0/2$, since $|\partial\Psi/\partial s - 1| = |\varphi(y)\beta'(|s|)| < \varepsilon_0/2$. The first term equals

$$\|d\Psi(\vec{v}_1)\| = \|\beta(|s|)\, d\varphi(\vec{v}_1)\| < \varepsilon_0/2 \; ,$$

as desired.

<u>5.13 Theorem</u>. Let M and N be manifolds of class C^p; let $f : M \to N$ be a C^r diffeomorphism $(1 \leq r \leq p \leq \infty)$. Given $\delta(x) > 0$, there is a C^p diffeomorphism $f_1 : M \to N$ which is a δ-approximation to f.

<u>Proof.</u> Let δ be small enough that if f_1 is a C^0 δ-approximation to f and $f_1(\text{Bd } M) \subset \text{Bd } N$, then f_1 is onto. Consider N as a submanifold of $D(N)$; then f is a C^r imbedding of M into $D(N)$. By Problem 3.7, there are positive functions δ_1 and δ_2 on M and $D(N)$ respectively, such that if g is a δ_1-approximation to f, and $h : D(N) \to D(N)$ is a δ_2-approximation to the identity, then hg is a δ-approximation to f.

Let $g : M \to D(N)$ be a C^p imbedding which is a δ_1-approximation to f. If δ_1 is small enough, g will carry $\text{Bd } M$ into the product neighborhood $\text{Bd } N \times (-1, 1)$ which is used to give $D(N)$ its differentiable structure. Then we may apply the preceding lemma, to obtain a C^p diffeomorphism h of $D(N)$ which carries $g(\text{Bd } M)$ onto $\text{Bd } N$. By making δ_1 small enough, we may make sure h is a δ_2-approximation to the identity.

Then $hg = f_1$ maps M into $D(N)$ and carries $\text{Bd } M$ into $\text{Bd } N$; a connectivity argument shows that M must be carried into the subset N of $D(N)$. Since f_1 is of class C^p, our result follows.

5.14 Problem.* Generalize the preceding theorem to prove that there is a C^r differentiable isotopy f_t between f and f_1, such that f_t is a diffeomorphism of M onto N for each t.

Hint: You will need to generalize Lemma 5.12 as follows: Let g_t be a C^r differentiable isotopy between two C^r imbeddings of M into $N \times R$ which is a δ-approximation to f for each t. Then there is a C^r differentiable isotopy h_t which is a diffeomorphism of $N \times R$ onto itself for each t, which satisfies (1) and (2) for each t, and in addition satisfies the conditions:

(3) If $g_t(M) \subset N$ for some t, then h_t is the identity map.

(4) If g_t is of class C^p for some t, so is h_t.

5.15 Problem*. Let $f : M \to N$ be a C^r map; N non-bounded. Prove that any sufficiently good strong approximation to f is differentiably homotopic to f.

Specifically, prove that given $\varepsilon(x) > 0$, there is a $\delta(x) > 0$ such that for any C^r map $g : M \to N$ which is a δ-approximation to f, there is a C^r differentiable homotopy f_t between f and g which is an ε-approximation to f, for each t.

Hint: Let N be a submanifold of euclidean space R^p; let r be a C^r retraction of a neighborhood of N in R^p, onto N. Deform f into g along straight lines in R^p, and then apply r to make the image remain in N.

5.16 Problem*. Generalize Problem 5.15 to the case where N has a boundary; draw the additional conclusion that whenever both $f(x)$ and $g(x)$ lie in Bd N, so does $f_t(x)$ for each t.

Hint: Imbed Bd N in R^p; extend to an imbedding $h : U \to R^{p+1}$, where U is a neighborhood of Bd N in $D(N)$, so that $\pi h(U) = h(N)$, where π is the projection of R^{p+1} onto R^p. Finally, choose an imbedding of $D(N)$ in some euclidean space which agrees with h in a neighborhood of Bd N, using 2.11. Then apply the techniques of 5.15, choosing the retraction r so that it carries a neighborhood of x in the plane normal to $D(N)$ at x onto x. This implies that r carries the line segment joining two nearby points of $h(\text{Bd } N)$ into $h(\text{Bd } N)$.

(a) Exercise*. Provide an alternate proof for Problem 5.14.

5.17 Problem*. A space X is locally contractible if for each point x of X and each neighborhood U of x, there is a neighborhood V of x such that the inclusion map $i : V \to U$ is homotopic to the constant map $c : V \to x$.

Let M and N be C^r manifolds; M compact. Prove that $F^r(M, N)$ is locally contractible in the C^1 topology. Prove also that the following spaces are locally contractible in the C^1 topology:

(1) The space of all C^r immersions of M into N.

(2) The space of all C^r imbeddings of M into N.

(3) The space of all C^r diffeomorphisms of M onto M.

§6. Uniqueness of the Double of a Manifold*.

6.1 Lemma*. Let M be a non-bounded C^r manifold; let W be a neighborhood of $M \times 0$ in $M \times R_+$, where $R_+ = [0, \infty)$. Let f be a C^r imbedding of W into $M \times R_+$ which equals the identity on $M \times 0$. There is a C^r diffeomorphism $\tilde{f}$ of W onto $f(W)$ which equals f in a neighborhood of the complement of W, and equals the identity in a neighborhood of $M \times 0$.

Proof. Let (x,t) denote the general point of $M \times R_+$. We may assume that W is the set of points (x,t) for which $0 \leq t < \beta(x)$, where $\beta(x) < 1$ is a positive C^r function on M. Let $f(x,t) = (X(x,t), T(x,t))$.

Step 1. Since f is non-singular, $\partial T(x,0)/\partial t$ is positive. Choose a positive C^r function $\epsilon(x) < 1$ on M such that

$$\partial T(x,t)/\partial t > 0 \qquad \text{for} \qquad 0 \leq t \leq \epsilon(x) \quad ,$$

and

$$1 > \epsilon(x)\,|\partial T(x,t)/\partial t| \qquad \text{for} \qquad 0 \leq t \leq \beta(x) \quad .$$

We define a diffeomorphism g of W with itself by the equation $g(x,t) = (x, \Psi(x,t))$, where

$$\Psi(x,t) = (1 - \alpha(t/\beta(x)))\,\epsilon(x) \cdot t + \alpha(t/\beta(x)) \cdot t \quad .$$

Here $\alpha(t)$ is, as usual, a monotonic C^∞ function which equals 0 for $t \leq 1/3$ and equals 1 for $2/3 \leq t$. The map g is a diffeomorphism, because $g(x,0) = (x,0)$, and $g(x,t) = (x,t)$ for $t \geq 2\beta(x)/3$, and

$$\partial \Psi(x,t)/\partial t = (1 - \alpha)\epsilon(x) + \alpha + (1 - \epsilon(x))(1/\beta(x)) \cdot \alpha' \geq \epsilon(x) > 0 \quad .$$

Let $f_1 = fg$. Then f_1 is a C^r diffeomorphism of W with $f(W)$, which equals f near $\bar{W} - W$. Furthermore, if we set $f_1(x,t) = (X_1, T_1)$, then

$$0 < \partial T_1(x,t)/\partial t < 1 \quad \text{for} \quad t < \beta(x)/3 \quad .$$

This follows from the fact that $T_1(x,t) = T(x, \epsilon(x)t)$ for $t < \beta(x)/3$, so that $\partial T_1(x,t)/\partial t = \epsilon(x) \cdot \partial T(x, \epsilon(x)t)/\partial t$, which is positive and less than 1, by choice of ϵ.

Step 2. Now we define a diffeomorphism h of W with itself by the equation $h(x,t) = (x, \varphi(x,t))$, where

$$\varphi(x,t) = \alpha(2t/\beta(x)) \cdot t + [1 - \alpha(2t/\beta(x))]\, T_1(x,t) \quad .$$

Again, h is a diffeomorphism because $\varphi(x,0) = 0$, and $\varphi(x,t) = t$ for

$t \geq \beta(x)/3$, and $\partial\varphi(x,t)/\partial t > 0$ for $t < \beta(x)/3$. This last inequality follows from the following computation:

$$\partial\varphi(x,t)/\partial t = \alpha + (1-\alpha)\cdot\partial T_1(x,t)/\partial t + [1 - T_1(x,t)/t]\cdot 2t\cdot\alpha'/\beta \quad ,$$

which is positive because $\partial T_1(x,t)/\partial t > 0$ and $T_1(x,t)/t = \partial T_1(x,t^*)/\partial t < 1$, where $0 < t^* < t$.

Let $f_2 = f_1 h^{-1}$. Then f_2 is a C^r diffeomorphism of W with $f(W)$ which equals f near $\bar{W} - W$. Furthermore, if we set $f_2(x,t) = (X_2,T_2)$, then $T_2(x,t) \equiv t$ in a neighborhood Y of $M\times 0$ in $M\times R_+$.

Step 3. If M is compact, the completion of the proof is easy. Choose δ small enough that $M\times[0,\delta]$ is contained in Y, and define $f_3(x,t) = (X_3,T_3)$, where $T_3 = T_2$ and

$$X_3(x,t) = X_2(x, \alpha(t/\delta)\cdot t) \quad .$$

f_3 is a diffeomorphism because the map $x \to X(x,t_0)$ is a diffeomorphism of M for $t_0 < \delta$, and $T_3(x,t) = t$ for $t < \delta$.

If M is not compact, more work is involved. We need the following lemma:

6.2 Lemma*. Let U be an open set in R^m whose closure is a ball. Let C be a compact subset of U; let V be a neighborhood of C whose closure is contained in U. Let f be a C^r imbedding of $U\times[0,a]$ into $R^m\times[0,a]$ which equals the identity on $U\times 0$, such that $f(x,t) = (X(x,t),t)$.

There is a C^r diffeomorphism $f_1 = (X_1,t)$ of $U\times[0,a]$ with $f(U\times[0,a])$ such that

(1) f_1 is the identity on $U\times 0$ and on $C\times[0,\delta]$, for some $\delta > 0$.

(2) $X_1(x,t) = X(x,t)$ outside $V\times[0,a/2]$.

(3) If f is the identity on some set $x\times[0,b]$, then f_1 is the identity on this set also.

Proof. Let $\varphi(x)$ be a C^∞ function on R^m which equals 1 on C and equals 0 outside V. Let $\gamma(t)$ be a monotonic C^∞ function on R which equals 1 for $t \leq 1/3$ and equals 0 for $2/3 \leq t$.

Let $f(x,t) = (X(x,t),t)$. Define $f_1(x,t) = (X_1(x,t),t)$, where

$$X_1(x,t) = X(x, t[1 - \gamma(t/\epsilon)\cdot\varphi(x)]) \quad .$$

Here $\epsilon < a/2$ is a positive number yet to be chosen.

Now for x in C and $t < \epsilon/3$, $X_1(x,t) = X(x,0) = x$, so condition (1) is satisfied. Condition (2) is clear, since $\varphi(x) = 0$ for x outside V, and $\gamma(t/\epsilon) = 0$ for $t > a/2$. Finally, suppose f is the identity on $x \times [0,b]$. If t is in $[0,b]$, so is $t(1 - \gamma\varphi)$, so that $X_1(x,t) = X(x,t(1 - \gamma\varphi)) = x$. Thus condition (3) holds.

All we need now to do is to show f is a diffeomorphism, and for this it will suffice that the map $h_t(x) = X_1(x,t)$ be an imbedding of $\bar{U}$ in R^m, for each t. By Exercise (b) of 3.10, there is a function $\delta(x) > 0$ on U such that if h_t is a δ-approximation to the map $g_t(x) = X(x,t)$, then h_t is an imbedding. Then there is a positive constant δ_0 such that h_t will be an imbedding if

$$|X_1(x,t) - X(x,t)| < \delta_0$$

and

$$|\partial X_1(x,t)/\partial x - \partial X(x,t)/\partial x| < \delta_0$$

for all x in $\bar{U}$.

By uniform continuity of X and $\partial X/\partial x$, there is a constant ϵ_0 such that $|X(x,t_0) - X(x,t)| < \delta_0$ and $|\partial X(x,t_0)/\partial x - \partial X(x,t)/\partial x| < \delta_0/2$ if x is in $\bar{U}$ and $|t - t_0| < \epsilon_0$. Let M be the maximum value of the entries of $|\partial X(x,t)/\partial t \cdot \partial\varphi(x)/\partial x|$ for x in $\bar{U}$. Choose ϵ to be the minimum of ϵ_0 and $\delta_0/2M$.

Now $X_1(x,t) = X(x,t_0)$, where $t_0 = t(1 - \gamma\varphi)$. Hence $|t - t_0| = |t \cdot \gamma(t/\epsilon) \cdot \varphi(x)| < \epsilon$, so that $|X_1(x,t) - X(x,t)| < \delta_0$, as desired. Also,

$$\partial X_1(x,t)/\partial x = \partial X(x,t_0)/\partial x \quad - \quad t\, \gamma(t/\epsilon)[\partial X(x,t_0)/\partial t \cdot \partial\varphi(x)/\partial x] \quad .$$

The first term is within $\delta_0/2$ of $\partial X(x,t)/\partial x$, and the second is within $\delta_0/2$ of zero. Hence our desired result follows.

<u>Completion of the proof of the theorem.</u> Cover M by a locally-finite collection of open sets U_i, for which there is a diffeomorphism h_i of $\bar{U}_i$ with a ball in R^m. Let C_i be a compact subset of U_i such that $\{C_i\}$ covers M. For convenience, assume U_1 and U_2 are empty.

Induction hypothesis: $f_i = (X_i, T_i)$ is a C^r diffeomorphism of Y with $f(Y)$ which equals f near $\bar{Y} - Y$, such that

(1) $T_i(x,t) = t$ on Y, and

(2) $X_i(x,t) = x$ for x in C_j and $0 \leq t \leq \delta_j$, for $j \leq i$.

Choose c so that $U_{i+1} \times [0,c)$ is contained in Y. By applying the lemma, we may obtain a diffeomorphism $f_{i+1} = (X_{i+1}, t)$ which equals f_i outside $U_{i+1} \times [0,c)$, such that $X_{i+1}(x,t) = x$ on $U_{i+1} \times 0$ and on $C_{i+1} \times [0,\delta_{i+1}]$, for some choice of δ_{i+1}. Because of condition (3) of the lemma, we will have $X_{i+1}(x,t) = x$ on $C_j \times [0,\delta_j]$ for all $j \leq i$.

We then define $\tilde{f}(x,t) = \lim_{i \to \infty} f_i(x,t)$ for (x,t) in Y; we extend it to U by letting it equal f_2 outside Y. It is easily seen that $\tilde{f}$ satisfies the requirements of the theorem.

<u>6.3 Theorem</u>*. The double of a C^r manifold M is uniquely determined, up to diffeomorphism.

<u>Proof</u>. Let $D(M)$ and $D'(M)$ be two differentiable manifolds obtained by using different product neighborhoods of Bd M to define the differentiable structure. Then $P : V \to \text{Bd } M \times (-1,1)$ and $P' : V' \to \text{Bd } M \times (-1,1)$ are diffeomorphisms of the open subsets V and V' of $D(M)$ and $D'(M)$, respectively, with Bd $M \times (-1,1)$. Now $P'P^{-1}$ maps a neighborhood W of Bd $M \times 0$ in Bd $M \times (-1,1)$ into Bd $M \times (-1,1)$; it equals the identity on Bd $M \times 0$, and is a diffeomorphism when restricted to the subsets $(\text{Bd } M \times [0,1)) \cap W = W_+$ and $(\text{Bd } M \times (-1,0]) \cap W = W_-$ of W. By the preceding theorem, there is a homeomorphism g of W with $P'P^{-1}(W)$ which equals $P'P^{-1}$ near Bd W, is a diffeomorphism when restricted to the subsets W_+ and W_-, and equals the identity in a neighborhood of Bd $M \times 0$.

Then $f = (P')^{-1} g P$ is defined on the neighborhood $P^{-1}(W)$ of Bd M in $D(M)$, and equals the identity map of $D(M) \to D'(M)$ near the boundary of this neighborhood. Hence f may be extended to a homeomorphism of $D(M)$ with $D'(M)$ by letting it equal the identity outside $P^{-1}(W)$. One checks readily that f is a diffeomorphism of $D(M)$ onto $D'(M)$.

(a) Exercise*. Let <u>Diff</u> M denote the space of all C^r diffeomorphisms of M onto itself. If f and g are in Diff M, f is said to be <u>weakly diffeotopic</u> to g if there is a C^r diffeomorphism F of $M \times R$ such that $F(x,t) = (f(x),t)$ in a neighborhood of $M \times (-\infty,0]$ and $F(x,t) = (g(x),t)$ in a neighborhood of $M \times [1,\infty)$. Show that this is an equivalence relation. Let $\tilde{\Gamma}(M)$ denote the equivalence classes. The composition of two diffeomorphisms is a diffeomorphism, so that Diff M is a group; show that this group operation makes $\tilde{\Gamma}(M)$ into a group.

(b) Exercise*. Show that if there exists a differentiable isotopy f_t between f and g which is a diffeomorphism for each t, then f and g are weakly diffeotopic. (The converse is an unsolved problem.)

(c) Exercise*. Let M be non-bounded; let G be a C^r diffeomorphism of $M \times I$ with itself such that $G(x,0) = (f(x),0)$ and $G(x,1) = (g(x),1)$. Prove that f and g are weakly diffeotopic.

(d) Exercise*. Let M be orientable; let $\Gamma(M)$ denote the subgroup of $\tilde{\Gamma}(M)$ generated by orientation-preserving diffeomorphisms. Show that $\tilde{\Gamma}(M)/\Gamma(M) \cong Z_2$ or 0 according as M possesses an orientation-reversing diffeomorphism or not.

(e) Exercise*. Show that $\Gamma(R^m) = 0$.

Hint: If f is a diffeomorphism of R^m, first deform f into the linear map which carries x into $Df(0) \cdot x$; then deform this into the identity map.

(f) Exercise*. If $f : M \to M$, the <u>support</u> of f is the closure of the set of all x for which $f(x) \neq x$. Let $\text{Diff}_c M$ denote those elements of Diff M having compact support; let $\tilde{\Gamma}_c(M)$ denote the weak diffeotopy classes of $\text{Diff}_c M$, where the support of the diffeotopy is required to have compact intersection with $M \times I$. If M is orientable, let $\Gamma_c(M)$ denote the subgroup generated by orientation-preserving diffeomorphisms. Milnor has proved that $\Gamma_c(R^6)$ is non-trivial (see [5] and [10]). Prove that in general $\Gamma_c(R^m)$ is abelian.

Hint: Given f and g, deform them so their supports are disjoint.

(g) Exercise*. Let f be a diffeomorphism of Bd M onto Bd N. Let $M \cup_f N$ denote the non-bounded manifold obtained from $M \cup N$ by identifying each x in Bd M with f(x). Put a differentiable structure on $M \cup_f N$ such that the inclusions of M and N are imbeddings. Show that the resulting differentiable manifold is unique up to diffeomorphism.

(h) Exercise*. Let f_0 be a fixed diffeomorphism of Bd M onto Bd N. For any diffeomorphism f of Bd M onto Bd N, let $\tilde{f}$ denote the element $f_0^{-1}f$ of Diff(Bd M). Show that up to diffeomorphism, the differentiable manifold $M \cup_f N$ depends only on the weak diffeotopy class of $\tilde{f}$.

CHAPTER II.

TRIANGULATIONS OF DIFFERENTIABLE MANIFOLDS

§7. Cell Complexes and Combinatorial Equivalence.

In this section we prove the theorem that two finite polyhedra in euclidean space may be subdivided into simplicial complexes in such a way that their intersection is a subcomplex of each of them (7.10). For this purpose, it is necessary first to define what is meant by a rectilinear cell complex, and to study some properties of cell complexes.

7.1 Definition. If $v_0, \dots, v_m$ are independent points of R^n, the simplex $\sigma = v_0 \dots v_m$ they span is the set of points x such that $x = \Sigma\, b_i v_i$, where $b_i \geq 0$ and $\Sigma\, b_i = 1$. The numbers b_i are called the barycentric coordinates of x. The point $\Sigma\, v_i/(m+1)$ is called the barycenter of σ and is denoted by $\tilde{\sigma}$. A face of a simplex σ is the simplex spanned by a subset of the vertices of σ. The simplex σ is homeomorphic to B^m; the interior of σ is called an open simplex.

If A and B are two subsets of R^n, the join $A * B$ of A and B is the union of all closed line segments joining a point of A and a point of B, providing no two of these line segments intersect except possibly at their end points. Then $\sigma = v_0 * (v_1 * (v_2 * \dots))$.

A (simplicial) complex K is a collection of simplices in R^n, such that

(1) Every face of a simplex of K is in K.

(2) The intersection of two simplices of K is a face of each of them.

(3) Each point of $|K|$ has a neighborhood intersecting only finitely many simplices of K.

Here $|K|$ denotes the union of the simplices of K and is called the polytope of K; sometimes it is called a polyhedron. (It would be more general to let the simplices of K lie in $R^\infty = \cup_n R^n$, but this definition will suffice for our purposes. We are in fact restricting ourselves to finite-dimensional complexes.)

A subdivision K' of K is a complex such that $|K'| = |K|$ and each simplex of K' is contained in a simplex of K. A subcomplex of K is a subset of K which is itself a complex.

If x is a point of $|K|$, the star of x in K is the union of the interiors of all simplices σ such that x lies in σ. It is an open subset of $|K|$, denoted by $St(x,K)$. If S is any subset of $|K|$, we define $St(S,K)$ to be the union of the sets $St(x,K)$ for all x in S. If J is a subcomplex of K, it is convenient to write $St(J,K)$ for $St(|J|,K)$.

Let σ be the simplex $v_0 \ldots v_m$. A map $f : \sigma \to R^p$ is linear if $f(x) = f(\Sigma\, b_i v_i) = \Sigma\, b_i\, f(v_i)$ for all x in σ. If K and L are complexes, a map $f : |K| \to |L|$ is linear relative to K and L if it carries each simplex of K linearly into a simplex of L. We often shorten this to the phrase — $f : K \to L$ is linear. The map $f : K \to L$ is piecewise-linear if for some subdivision K' of K, $f : K' \to L$ is linear. $f : K \to L$ is a linear isomorphism if it is a homeomorphism of $|K|$ onto $|L|$ and carries each simplex of K linearly onto one of L.

It was long a famous unsolved problem (often called the Hauptvermutung) whether the existence of a homeomorphism between $|K|$ and $|L|$ implied the existence of a piecewise-linear homeomorphism of $|K|$ onto $|L|$. Partial answers were: Yes, if dimension $K = 2$ [11]; yes, if $|K|$ is a 3-manifold [1, 7]; yes, if K and L are smooth triangulations of diffeomorphic manifolds (this we shall prove). Recently, Milnor has shown that the answer is no if dimension $K \geq 6$ [6]. It is still unknown if the answer is yes when $|K|$ is a manifold.

(a) Exercise. Let K be a complex. Show that the closure of $St(x,K)$, denoted by $\overline{St}(x,K)$, is the polytope of a finite subcomplex of K. We sometimes use $\overline{St}(x,K)$ to denote the complex as well as the polytope, where no confusion will arise.

(b) Exercise. Show that if $f : K \to L$ is a piecewise-linear homeomorphism of $|K|$ onto $|L|$, then they have subdivisions K' and L' such that $f : K' \to L'$ is a linear isomorphism.

7.2 Definition. Let c be a bounded subset of R^n, consisting of all points x satisfying a system of linear equations and linear inequalities

$$L_i(x) = \Sigma_j \, a_{ij} \, x^j \geq b_i \; ; \quad i = 1, \ldots, p \; ;$$

such a set c is called a (rectilinear) cell. It is a compact convex subset of R^n. (Convexity of a set c means that c contains each line segment joining two of its points.)

The dimension m of c is the dimension of the smallest dimensional plane $\mathcal{P}$ containing c; this means that c contains $m + 1$ independent points, but not $m + 2$. Since c must contain the simplex spanned by these $m + 1$ points, it must have interior points as a subset of $\mathcal{P}$. Let Int c denote these points; let Bd c be the remainder of c. We show there is a homeomorphism of c with the m-ball B^m carrying Bd c onto S^{m-1}:

Let us adjoin to the system defining c a set of equations for $\mathcal{P}$. Some of the inequalities of the system may now be redundant; let $L_i(x) \geq b_i$, $i = 1, \ldots, p$, be a minimal system of inequalities which, along with the equations for $\mathcal{P}$, serve to determine c. We note that for $i = 1, \ldots, p$, the hyperplane $L_i(x) = b_i$ intersects $\mathcal{P}$ in a plane of dimension $m-1$; for otherwise, this hyperplane would contain $\mathcal{P}$ and the corresponding inequality could be discarded without changing the set c. Furthermore, Int c equals the set A of those points for which each of these inequalities is strict: Clearly, A is contained in Int c. On the other hand, if x_0 is a point of the intersection of $\mathcal{P}$ with the hyperplane $L_j(x) = b_j$, then arbitrarily near x_0 are points of $\mathcal{P}$ for which $L_j(x) < b_j$, so that x_0 does not lie in Int c.

Let $\tilde{c}$ be a point of Int c and let r be a ray in $\mathcal{P}$ beginning at $\tilde{c}$. The intersection of r with c is non-degenerate, compact, and convex; i.e., a closed interval. One end point is $\tilde{c}$; the other is necessarily a point y of Bd c. Each point x of the open line segment

$\tilde{c}y$ necessarily lies in Int c, for $L_i(\tilde{c}) > b_i$ and $L_i(y) \geq b_i$ for $i = 1, \ldots, p$, so that $L_i(x) > b_i$. Hence c is equal to the join of $\tilde{c}$ with Bd c.

To complete the proof we need only show Bd c homeomorphic with S^{m-1}. Without loss of generality we may assume $\mathcal{P}$ is R^m and $\tilde{c}$ is the origin. The map $x \to x/\|x\|$ carries $R^m - 0$ continuously onto the unit sphere; it is necessarily a homeomorphism when restricted to Bd c.

<u>7.3 Lemma</u>. Let c be an m-cell. Then Bd c is the union of a finite number of m-1 cells, each the intersection of an m-1 plane with Bd c. These cells are uniquely determined by c.

<u>Proof</u>. Let $\mathcal{P}$ be the m-plane containing c; let c be given by equations for $\mathcal{P}$, along with a <u>minimal</u> set of inequalities $L_i(x) \geq b_i$, $i = 1, \ldots, p$.

Let d_i denote the set of points x of c for which $L_i(x) = b_i$; then d_i is a cell, and Bd c is the union of these cells. Furthermore, the intersection of the hyperplane $L_i(x) = b_i$ with $\mathcal{P}$ is an $m - 1$ plane $\mathcal{P}_i$, whose intersection with Bd c is precisely d_i.

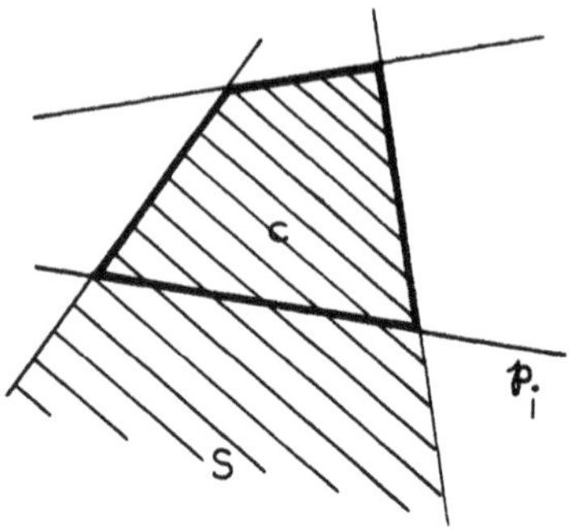

We prove that d_i is a cell of dimension $m - 1$. Let S denote the subset of $\mathcal{P}$ for which $L_j(x) > b_j$ for all $j \neq i$. Then S is convex and contains Int c; in particular, it contains a point x for which $L_i(x) > b_i$. It also contains a point y such that $L_i(y) < b_i$, since otherwise $\bar{S}$ would lie entirely in the region $L_i(x) \geq b_i$, so that discarding the inequality $L_i(x) \geq b_i$ from the set of inequalities for c would not change the set c. By convexity, S contains a point z such that $L_i(z) = b_i$. Then $S \cap \mathcal{P}_i$ is not empty; since S is open in $\mathcal{P}$, $S \cap \mathcal{P}_i$ is open in $\mathcal{P}_i$. Since $S \cap \mathcal{P}_i \subset d_i$, d_i must be a cell of dimension $m - 1$.

The uniqueness of the cells d_i is easy: Bd c is contained in the union of p planes of dimension m-1. Hence the intersection of any other m-1 plane with Bd c lies in the union of finitely many planes of dimension less than m-1, and hence has dimension less than m-1.

<u>7.4 Definition</u>. If c is an m-cell, each of the m - 1 cells d_i into which Bd c decomposes is called a <u>face</u> of c; each m - 2 face of a d_i is called a m - 2 <u>face</u> of c. And so on. By convention, the empty set is also a face of c, as is c itself.

<u>7.5 Lemma</u>. Let c be a cell given by a system of linear equations and inequalities. Replacing some inequalities by equalities determines a face of c, and conversely.

<u>Proof</u>. We proceed by induction on m, the dimension of c. If m = 0, the lemma is trivial. If m > 0, let $\mathcal{P}$ be the m-plane containing c; adjoin to the system determining c a set of equations for $\mathcal{P}$. This we clearly may do without loss of generality. Let $L_i(x) \geq b_i$ be the inequalities for c, so numbered that those for which $1 \leq i \leq p$ form a minimal set. Then let us replace some inequality $L_j(x) \geq b_j$ by an equality. If $j \leq p$, this determines an m-1 face d_j of c, as we have just proved.

In the case $j > p$, consider the subset of c satisfying $L_j(x) = b_j$. If the hyperplane $L_j(x) = b_j$ contains $\mathcal{P}$, this gives the trivial face c; if it does not intersect c, it gives the empty face. Otherwise, let e be the intersection of c with this hyperplane; it is a cell. Now the intersection of $\mathcal{P}$ with this hyperplane is an m - 1 plane, so that arbitrarily near each point of e are points y of $\mathcal{P}$ for which $L_j(y) < b_j$. Hence e lies on the boundary of c.

Since e is convex, it must lie in some m-1 face d_i: otherwise, let q be the smallest integer such that e lies in $d_1 \cup \ldots \cup d_q$. Let x be a point of e not in $d_1 \cup \ldots \cup d_{q-1}$; let y be a point of e not in d_q. Then

$$L_i(x) > b_i \quad \text{for} \quad i < q \quad \text{and} \quad L_q(x) = b_q \; ;$$
$$L_i(y) \geq b_i \quad \text{for} \quad i < q \quad \text{and} \quad L_q(y) > b_q \; .$$

The point $z = (x + y)/2$ lies in e, but $L_i(z) > b_i$ for all $i \leq q$, so that z does not lie in $d_1 \cup \ldots \cup d_q$, contrary to hypothesis.

The system of equations and inequalities for c, with the equation $L_i(x) = b_i$ adjoined, determine the $m - 1$ cell d_i which contains e. Furthermore, e is obtained from this system by replacing $L_j(x) \geq b_j$ by an equality. By the induction hypothesis, e is a face of d_i, and of c.

7.6 Definition. A (rectilinear) cell complex K is a collection of cells in R^n such that

(1) Each face of a cell in K is in K.

(2) The intersection of two cells of K is a face of each of them.

(3) Each point of $|K|$ has a neighborhood intersecting only finitely many cells of K. (Here $|K|$ denotes the union of the cells of K.)

In order that this definition be non-vacuous, we need to note that a single cell c, along with its faces, constitutes a cell complex. This follows from the preceding lemma (see Exercise (a) below).

A subdivision of a cell complex K is a cell complex K' such that $|K'| = |K|$ and each cell of K' is contained in one of K. The dimension of K is the dimension of the largest dimensional cell in K. The p-skeleton of K, denoted by K^p, is the collection of all cells of K having dimension at most p. One checks that K^p is a subcomplex of K.

(a) Exercise. If K consists of a single cell, along with its faces, show that K is a cell complex.

(b) Exercise. Show that the simplex $\sigma = v_0 \ldots v_m$ is an m-cell, and that the notion of "face" is the same whether one uses the definition in 7.1 or the one in 7.4. Show that a simplicial complex is a special kind of cell complex.

7.7 Lemma. Let K_1 and K_2 be cell complexes such that $|K_1| = |K_2|$. Then K_1 and K_2 have a common subdivision.

Proof. It is clear that the intersection of two cells is again a cell. Let L be the collection of all cells of the form $c_1 \cap c_2$, where c_1 is a cell of K_1 and c_2 is a cell of K_2. Then L is a cell complex, and $|K_1| = |K_2| = |L|$.

(a) Exercise. Show that any face of $c_1 \cap c_2$ is of the form $e_1 \cap e_2$, where e_i is a face of c_i; and conversely.

(b) Exercise. Show that L is a cell complex.

7.8 Lemma. Any cell complex K has a simplicial subdivision.

Proof. We proceed by induction on the dimension m of K. If $m = 0$ or $m = 1$, K is already a simplicial complex. In general, suppose L is a simplicial subdivision of K^{m-1}, the $m - 1$ skeleton of K. If c is an m-cell of K, let $\sigma_1,\ldots,\sigma_p$ be the simplices of L lying in Bd c. Choose an interior point $\tilde{c}$ of c; adjoin to the collection L the simplices $\sigma_1 * \tilde{c},\ldots,\sigma_p * \tilde{c}$. (Recall that $\sigma_i * \tilde{c}$ denotes the join of σ_i and $\tilde{c}$.) If we carry out this construction for each m-cell c of K, the result will be a simplicial complex K' which is a subdivision of K.

It is often convenient to choose $\tilde{c}$ as the centroid of c, in order that K' be canonically defined. If K is already simplicial, the centroid is the same as the barycenter, and this subdivision K' is called the <u>barycentric subdivision</u> of K.

(a) Exercise. Check that K' is a complex, and that it is a subdivision of K.

7.9 Corollary. If K_1 and K_2 are simplicial complexes such that $|K_1| = |K_2|$, K_1 and K_2 have a common simplicial subdivision.

7.10 Theorem. Let K_1 and K_2 be two finite simplicial complexes in R^n. There are simplicial subdivisions K_1' and K_2' of K_1 and K_2, respectively, such that $K_1' \cup K_2'$ is a simplicial complex.

Proof. A rectilinear triangulation of R^n is a simplicial complex L such that $|L| = R^n$. We prove that some subdivision of K_1 is a subcomplex of a rectilinear triangulation L_1 of R^n:

Let $\sigma_1,\ldots,\sigma_p$ be the simplices of K_1. Choose a rectilinear triangulation J_1 of R^n which contains σ_1. One way to construct such a J_1 is to take any rectilinear triangulation J of R^n and find a non-singular affine transformation h_1 of R^n which carries one of its simplices onto σ_1; the collection $h_1(J)$ will be the required complex. (An affine transformation is a linear transformation composed with a translation.) Similarly, let J_i be a rectilinear triangulation of R^n containing σ_i; let L_1 be a common subdivision of $J_1,\ldots,J_p$ (using 7.9). Then $|K_1|$ is the polytope of a subcomplex of L_1, and this subcomplex is a subdivision of K_1.

Similarly, let L_2 be a rectilinear triangulation of R^n containing some subdivision of K_2 as a subcomplex. Let L be a common subdivision of L_1 and L_2, and let K_1' and K_2' be the subcomplexes of L having polytopes $|K_1|$ and $|K_2|$, respectively.

7.11 Problem*. Generalize this theorem to the case in which K_1 and K_2 are not finite. You will need some further hypothesis to avoid the case where $|K_1|$ is the set $[0,1]\times 0$ in R^2 and $|K_2|$ is the union of the sets $[0,1]\times 1/n$ for $n = 1,2,\ldots$.

7.12 Definition. The proof of Lemma 7.8 generalizes to the following situation: Let K_1 be a subcomplex of the simplicial complex K; let K_1' be a simplicial subdivision of K_1. We define a canonical way of extending K_1' to a subdivision of K, without subdividing any simplex outside $\mathrm{St}(K_1,K)$. We will call it the standard extension of K_1' to a subdivision of K:

Every simplex of K_1' belongs to K', of course, as does every simplex of K which is outside $\mathrm{St}(K_1,K)$. The simplices whose interiors lie in $A = \mathrm{St}(K_1,K) - |K_1|$ are subdivided step-by-step as follows: No vertices of K lie in A. Each 1-simplex σ of K whose interior lies in A

is subdivided into two 1-simplices, the barycenter of σ being the extra vertex. In general, suppose the $m - 1$ simplices have already been subdivided. For any m-simplex σ, its boundary has already been subdivided. Let $\tilde{\sigma}$ be its barycenter, and for each simplex s of the subdivision of Bd σ, let $s * \tilde{\sigma}$ be a simplex of the subdivision of σ. (Convention: the join of the empty set with $\tilde{\sigma}$ is $\tilde{\sigma}$ itself.) After a finite number of steps, we will have the required subdivision of K.

(a) Exercise. Check that the collection of simplices obtained in this way is a complex whose polytope is $|K|$.

7.13 Problem. Let K be a simplicial complex; let $\{A_i\}$ be a locally-finite collection of subsets of $|K|$. Let $K_1, K_2, \ldots$ be a sequence of simplicial subdivisions of K such that K_{i+1} equals K_i outside A_i. This means that any simplex of K_i which does not intersect A_i is a simplex of K_{i+1} as well. Let $\lim_{i \to \infty} K_i$ denote the collection of those simplices σ such that σ belongs to K_i for all i greater than some integer N_σ. Show that this collection is a subdivision of K.

(a) Exercise. Let K be a simplicial complex; let $\delta(x)$ be a positive continuous function on $|K|$. There is a subdivision K' of K such that for any simplex σ of K', the diameter of σ is less than the minimum of $\delta(x)$, for x in σ.

§8. Immersions and Imbeddings of Complexes

In this section, we define the notion of a C^r map $f : K \to M$, where K is a simplicial complex and M is a differentiable manifold, and we develop a theory of such maps analogous to that for maps of one manifold into another. In particular, we define (8.2) the differential of such a map and use this to define (8.3) the concepts of immersion, imbedding, and triangulation (which is the analogue of diffeomorphism). As before we define (8.5) what is meant by a strong C^1 approximation to a map $f : K \to M$, and prove the fundamental theorem which states that a sufficiently good strong C^1 approximation g to an immersion or imbedding is also an immersion or imbedding, respectively. (The theorem also holds for triangulations, with the additional hypothesis that g carries $\text{Bd } |K|$ into $\text{Bd } M$.)

From now on, we restrict ourselves to simplicial complexes and subdivisions, unless otherwise specified. The integer r $(1 \leq r \leq \infty)$ will remain fixed, for the remainder of this chapter.

<u>8.1 Definition</u>. Let K be a complex. The map $f : |K| \to M$ is <u>differentiable of class C^r relative to K</u> if $f|\sigma$ is of class C^r, for each simplex σ of K. We usually shorten this to the phrase, $f : K \to M$ is of class C^r. The map f is said to be <u>non-degenerate</u> if $f|\sigma$ has rank equal to the dimension of σ, for each σ in K.

We wish to generalize the notions of immersion and imbedding to this situation. As an analogue to a C^r imbedding of a manifold, one might take a C^r non-degenerate homeomorphism of a complex. That is, until one

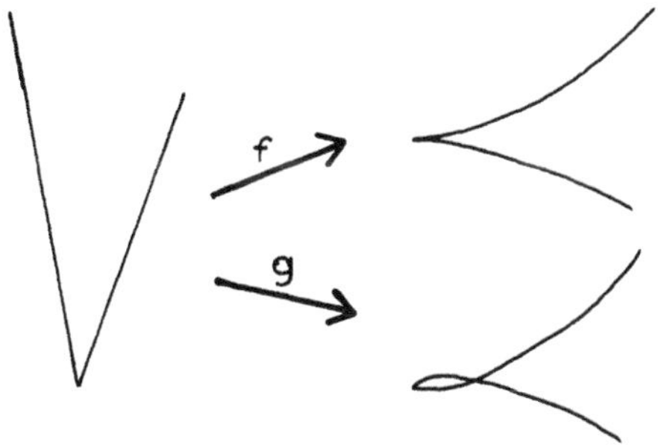

looks at the homeomorphism f of the accompanying illustration. The crucial property of imbeddings — that any sufficiently good strong C^1 approximation to an imbedding is also an imbedding — fails here, since arbitrarily close to f are maps like g. This example shows us that we must seek further to find the proper generalization.

8.2 Definition. Let $f : \sigma \to R^n$ be a C^r map. Given the point b of σ, define the map $df_b : \sigma \to R^n$ by the equation

$$df_b(x) = Df(b) \cdot (x - b) \quad .$$

Here x and b are written as column matrices, as usual; and we choose some orthonormal coordinate system in the plane in which σ lies, in order to compute Df. The map df_b is independent of this choice of coordinates in the plane of σ:

Let R be some ray in σ beginning at b. Then $f|R$ is a C^r curve in R^n, which we suppose parametrized by arc length along R. Now $df_b(x)$ is merely the tangent vector of this curve at $f(b)$, multiplied by $\| x - b \|$. Hence $df_b(x)$ does not depend on the choice of coordinates, since it involves only the distance function in the plane of σ. Furthermore, it follows that $df_b|R$ depends only on $f|R$, not on any other values of f.

Now if $f : K \to R^n$ is a C^r map, we have maps $df_b : \sigma \to R^n$ defined for each σ in $\overline{St}(b,K)$. These maps agree on the intersection of any two simplices of $\overline{St}(b,K)$, since either (1) one is a face of the other, or (2) their intersection lies in the union of rays emanating from b. Hence the map

$$df_b : \overline{St}(b,K) \to R^n$$

is well-defined and continuous. By analogy with the situation for differentiable manifolds, we call it the differential of f.

8.3 Definition. Let $f : K \to M$ be a C^r map, where M is a C^r submanifold of R^n. The map f is said to be an immersion if

$$df_b : \overline{St}(b,K) \to R^n$$

is one-to-one for each b. An immersion which is a homeomorphism is called n imbedding; if it is also a homeomorphism onto, it is called a ^{r} triangulation of M.

If x and b belong to the simplex σ of K, then the requirement that $df_b|\sigma$ be one-to-one implies that the matrix $D(f|\sigma)$ has rank at equal to the dimension of σ. Hence an immersion is automatically a non-egenerate map. The converse is not true in general, as the example in .1 shows. However, in the case where f is a homeomorphism onto, there is converse, as the following theorem shows.

(a) Exercise. Let $f : K \to M$ be a C^r immersion of the complex in the manifold M; let $g : M \to N$ be a C^r immersion of M in the anifold N. Show that gf is a C^r immersion of the complex K in N.

(b) Exercise. Show that the definitions of C^r immersion, imbedding, and triangulation are independent of which particular imbedding of M n euclidean space is chosen.

8.4 Theorem. Let $f : K \to M$ be a C^r non-degenerate map which is a homeomorphism of $|K|$ onto M. Then f is a C^r triangulation of M.

Proof. We need to prove that df_b is 1-1 on $\overline{St}(b,K)$. The only way it could fail to be 1-1 is for it to map rays R_1 and R_2 beginning at b into a single ray in R^n beginning at f(b). This means that $f|R_1$ and $f|R_2$ are C^r curves in M which are tangent to each other at f(b). Since f is non-degenerate, the restriction of df_b to any simplex of $\overline{St}(b,K)$ is 1-1. Hence no neighborhoods of b in R_1 and R_2 lie in the same simplex of $\overline{St}(b, k)$.

Choose a coordinate system (U,h) in M about f(b). Let x_n and y_n be sequences of points of R_1 and R_2, respectively, converging to b; let x_n', y_n', b' be their respective images under hf. Let s be the simplex of $\overline{St}(b,K)$ into whose interior R_1 extends; let V be the open set $St(\tilde{s},K) \cap f^{-1}(U)$. Now hf(V) is open in R^m; it contains x_n', but not y_n'. If n is large, x_n lies in V, so the line segment $x_n'y_n'$ intersects $hf(\bar{V} - V)$ in at least one point z_n'; further, $z_n' \neq b'$ because the angle between $x_n' - b'$ and $y_n' - b'$ approaches 0, not π, as $n \to \infty$.

Let z_n denote the point of $\bar{V} - V$ corresponding to z_n'. By passing to subsequences and renumbering, we may assume z_n lies in a single simplex of K and that the unit vectors $(z_n - b)/\| z_n - b \|$ converge, say to the vector $\vec{u}$. Let R_3 be the ray along which $\vec{u}$ lies.

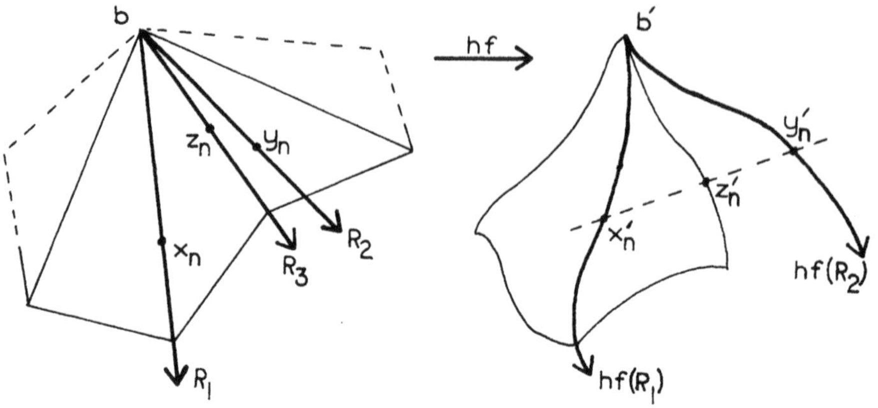

Now R_1 and R_3 lie in a single closed simplex of $\overline{St}(b,K)$, although $hf|R_1$ and $hf|R_3$ are tangent at b'; this contradicts the assumption that f is non-degenerate. The computation is as follows: $(x_n' - b')/\| x_n' - b' \|$ approaches the unit tangent vector to $hf(R_1)$ at b'. Further,

$$\frac{z_n' - b'}{\| z_n - b \|} = D(hf)(b) \cdot \frac{(z_n - b)}{\| z_n - b \|} + \frac{o(\| z_n - b \|)}{\| z_n - b \|} .$$

Since the angle between $x_n' - b'$ and $z_n' - b'$ goes to zero, the left side of this equation approaches a vector tangent to $hf|R_1$ at b'. The right side approaches $D(hf)(b) \cdot \vec{u}$, which is tangent to $hf|R_3$ at b'.

(The picture is, unfortunately, a bit misleading, since it shows z_n lying <u>on</u> R_3, which we can't guarantee. The proof would be easier if this were the case.)

(a) Exercise*. Assume the hypotheses of 8.4. Show that K is a <u>combinatorial</u> <u>n-manifold</u>. This means that for each k-simplex s of K, the <u>link</u> of s, $\text{Lk } s = \overline{St}\ \tilde{s} - St\ s$, is a combinatorial n-k-1 sphere; that is, it has a subdivision isomorphic with a subdivision of the boundary of an n-k simplex σ.

Hint: Prove it first when $s = v$, a vertex of K. Find a linear isomorphism h of $\overline{St}\, v$ with a complex L in R^n, such that $h(v) = \tilde{\sigma}$ and $|L| \supset \sigma$. Find a subdivision σ' of σ such that the inclusion $i : \sigma' \to L$ is linear. Radial projection from $\tilde{\sigma}$ of Bd σ' onto Bd $|L|$ will then be a linear homeomorphism.

<u>8.5 Definition</u>. Let $f : K \to R^n$ be a C^r map. Let δ be a positive continuous function on K. By analogy with 3.5, the map $g : |K| \to R^n$ is said to be a δ-<u>approximation</u> to f if

(1) For some subdivision K' of K, $g : K' \to R^n$ is a C^r map.

(2) $\| f(b) - g(b) \| < \delta(b)$ for all b in $|K|$.

(3) $\| df_b(x) - dg_b(x) \| \leq \delta(b) \| x - b \|$ for all b in $|K|$ and all x in $\overline{St}(b, K')$.

Such a map g is also called a <u>strong</u> C^1 <u>approximation</u> to f.

Note that this definition is independent of coordinate systems, depending only on the distance functions $\| x - y \|$ in the euclidean space containing K, and in R^n. Note also that if K is finite, δ may be required to be constant, without changing the situation; we shall always assume δ is constant in this case.

As in 3.6, condition (3) holds if the Jacobians of f and g are close to each other, since $\| df_b(x) - dg_b(x) \| / \| x - b \| \leq \sqrt{n}\, p\, |D(f|\sigma)(b) - D(g|\sigma)(b)|$, where p equals the dimension of the simplex σ containing x.

Let $f : K \to R^n$ and $g : K' \to R^n$ be C^r maps. If U is an open subset of $|K|$, g is said to be a δ-<u>approximation to</u> f <u>on</u> U if conditions (2) and (3) hold for all b in U. Supposing f and g to be non-degenerate on U, g is said to be an α <u>angle-approximation to</u> f <u>on</u> U if the angle between the vectors $df_b(x)$ and $dg_b(x)$ is less than α, for all b in U and all $x \neq b$ in $\overline{St}(b, K')$.

<u>8.6 Lemma</u>. Let $f : K \to R^n$ be a non-degenerate C^r map. Let $\alpha > 0$. If b is in $|K|$, there is a neighborhood of b on which df_b is an α angle-approximation to f.

Proof. It will suffice to prove this in the case where K is a single simplex σ. If x is a point of σ, and $\vec{u}$ is a unit vector in the plane of σ, we wish to make the angle between $Df(x)\cdot\vec{u}$ and $D(df_b)(x)\cdot\vec{u}$ less than α. But $df_b(x) = Df(b)\cdot(x-b)$, so $D(df_b)(x) = Df(b)$. Now the angle between $Df(x)\cdot\vec{u}$ and $Df(b)\cdot\vec{u}$ is a continuous function of x and $\vec{u}$, and its value is zero when $x = b$. Since the domain of $\vec{u}$ is a compact set, there is some neighborhood U of b such that this angle is less than α when x is in U, for any $\vec{u}$.

8.7 Lemma. Let $f : K \to R^n$ be a non-degenerate C^r map; K finite. Given $\alpha > 0$, there is a $\delta > 0$ such that any non-degenerate C^r map $g : K' \to R^n$ which is a δ-approximation to f on U is an α angle-approximation to f on U, for any U.

Proof. Let $\epsilon = \min \| df_b(x) \| / \| x - b \|$ for all b in $|K|$ and all $x \neq b$ in $\overline{St}(b,K)$; since f is non-degenerate, $\epsilon > 0$. Let $\delta = \alpha\epsilon/\pi$; let $g : K' \to R^n$ be a non-degenerate C^r map which is a δ-approximation to f on U.

Let σ be a simplex of K'; let b and x lie in σ; let b lie in U. The angle θ between $df_b(x)$ and $dg_b(x)$ is the same as that between $\vec{v} = D(f|\sigma)(b)\cdot\vec{u}$ and $\vec{w} = D(g|\sigma)(b)\cdot\vec{u}$, where $\vec{u} = (x-b)/\|x-b\|$. The angle between $\vec{v}$ and $\vec{w}$ is less than $\| \vec{v} - \vec{w} \| \pi/\|\vec{v}\|$, which is less than $\delta\pi/\epsilon = \alpha$. (For if θ is acute, then $\| \vec{v} - \vec{w} \| \geq \|\vec{v}\| \sin\theta \geq \|\vec{v}\| 2\theta/\pi$; and if θ is not acute, $\| \vec{v} - \vec{w} \| > \|\vec{v}\| \geq \|\vec{v}\| \theta/\pi$.)

8.8 Theorem. Let $f : K \to R^n$ be a C^r immersion, or imbedding; let K be finite. There is a $\delta > 0$ such that any δ-approximation to f is an immersion, or imbedding, respectively.

Proof. We first make two preliminary definitions, and prove a special case of the theorem.

(1) Let σ_1 and σ_2 be two simplices, with $\sigma = \sigma_1 \cap \sigma_2$ a proper face of each. The angle from σ_1 to σ_2 is defined as follows: Let $\tilde{\sigma}$

be the barycenter of σ. Consider the angle $\theta(y,z)$ between the line segments $y\tilde{\sigma}$ and $z\tilde{\sigma}$, where y lies in the face of σ_1 opposite σ and z is any point of σ_2 different from $\tilde{\sigma}$. This angle is positive. The minimum such angle θ is called the angle from σ_1 and σ_2.

Now θ is positive: If we extend the ray from $\tilde{\sigma}$ to z, it will intersect $\sigma_2 - \mathrm{St}(\tilde{\sigma})$. So we would get the same minimum if z ranged only over $\sigma_2 - \mathrm{St}(\tilde{\sigma})$; since y and z then range over compact sets, θ is positive.

Now the angle θ from σ_1 to σ_2 is a function of the vertices of σ_1 and σ_2. If we vary these vertices continuously (without allowing σ_1 and σ_2 to become degenerate), then θ varies continuously as well (see Exercise (a)).

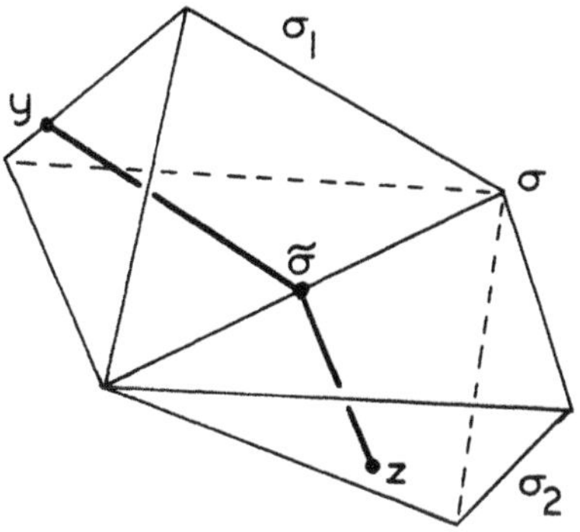

(2) Let σ be a proper face of the simplex σ_1. We define the <u>projection</u> of σ_1 onto σ as follows: First, it is the identity on σ itself. Second, suppose p belongs to $\sigma_1 - \sigma$. Let s_1 be the face of σ_1 opposite σ; let $\tilde{\sigma}$ be the barycenter of σ. There are uniquely defined points y of s_1 and x of σ such that the line segments $y\tilde{\sigma}$ and px are parallel. The projection of σ_1 onto σ is defined to carry p into x.

Projection is a continuous map of σ_1 onto σ; furthermore, it is natural in the sense that it is invariant under a linear isomorphism of σ_1 with another simplex (see Exercise (b)).

The fact about projection which is useful for us is the following: Let $\sigma = \sigma_1 \cap \sigma_2$ be a proper face of both σ_1 and σ_2; let p belong to

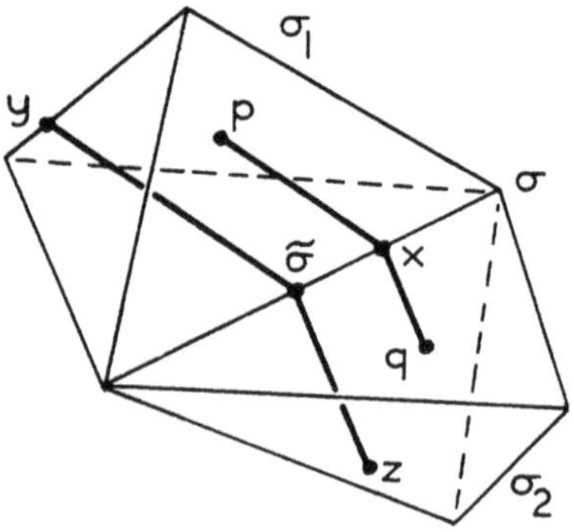

$\sigma_1 - \sigma$ and q to $\sigma_2 - \sigma$. If the projection of σ_1 onto σ carries p into x, then the angle between the line segments px and xq is no less than the angle from σ_1 to σ_2. To prove this, merely choose a point z in σ_2 such that $\tilde{\sigma}z$ is parallel to xq, and use the definition of the angle from σ_1 to σ_2.

(3) Now we prove a very special case of the theorem: Let L be a complex consisting of two line segments e_1 and e_2 with vertex v in common, and let $F : L \to R^n$ be a 1-1 linear map. Let the images under F of the two line segments make an angle of $\alpha > 0$ with each other. If $G : L' \to R^n$ is a non-degenerate C^r map which is an $\alpha/2$ angle-approximation to F, it follows that G is 1-1.

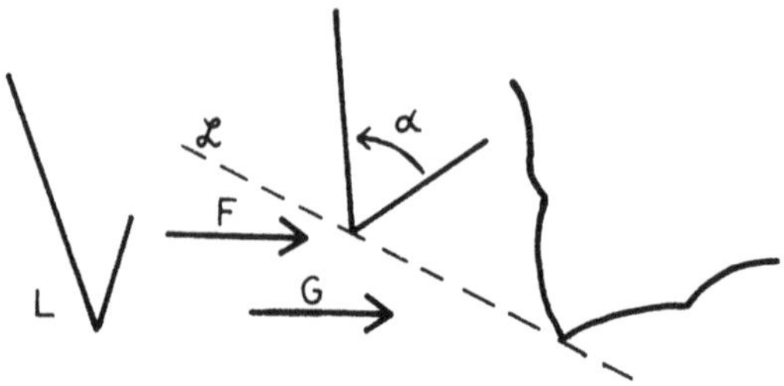

For let $\mathcal{P}$ be a 2-plane containing F(L). ($\mathcal{P}$ is in fact unique unless $\alpha = \pi$.) Let $\mathcal{L}$ be the straight line in $\mathcal{P}$ perpendicular to the bisector of α. Now no vector $dG_b(x)$ is perpendicular to $\mathcal{L}$, so each has a non-zero projection on $\mathcal{L}$. If we let π denote the orthogonal projection of R^n onto $\mathcal{L}$, it follows that πG is 1-1. Hence G is 1-1.

This special case contains the crux of the entire argument.

Proof of the theorem.

(4) Let f be an immersion. First, choose δ small enough that any δ-approximation to f is a non-degenerate map. This is easily done.

A second condition on δ: For each pair of simplices σ_1 and σ_2 of K such that $\sigma = \sigma_1 \cap \sigma_2$ is a proper face of each, consider

$$df_b : \sigma_1 \cup \sigma_2 \to R^n ,$$

where b is any point of σ. It is a 1-1 linear map, so there is a positive angle from the image of σ_1 to the image of σ_2. This angle varies continuously with b, so there is a minimum such angle $\alpha(\sigma_1,\sigma_2)$, as b ranges over σ. Let α be the minimum of $\alpha(\sigma_1,\sigma_2)$ for all such pairs σ_1,σ_2. Choose δ small enough that any δ-approximation g to f is an $\alpha/6$ angle-approximation to f (using 8.7).

It follows that any δ-approximation $g : K' \to R^n$ to f is an immersion, as we now prove.

Let b be a point of $|K|$; we need to prove

$$dg_b : \overline{St}(b,K') \to R^n$$

is 1-1. Suppose $dg_b(p) = dg_b(q)$, for some points p and q in $\overline{St}(b,K')$. Then dg_b maps the two rays beginning at b and passing through p and q, respectively, onto a single ray of R^n. Hence there are points p and q on these rays as close to b as we wish such that $dg_b(p) = dg_b(q)$.

There is a neighborhood U of b such that U lies in $St(b,K')$, such that $df_b(x)$ is an $\alpha/6$ angle-approximation to $f(x)$ for x in U, and such that $dg_b(x)$ is an $\alpha/6$ angle-approximation to $g(x)$ for x in U (by 8.6). Then on the set U, $dg_b(x)$ is an $\alpha/2$ angle-approximation to $df_b(x)$.

If p and q lie in the same simplex σ of K, take them close enough to b that the line segment pq lies in U. Let x be the midpoint of pq; let $e_1 = px$ and $e_2 = xq$; let $L = e_1 \cup e_2$. Now df_b is linear on the complex L, and $df_b(e_1)$ and $df_b(e_2)$ make an angle of π with each other. Further, dg_b is an $\alpha/2$ angle-approximation to df_b on L. Since $\alpha \leq \pi$, the special case of our theorem applies to show that dg_b is 1-1 on L, contrary to hypothesis.

Otherwise, let σ_1 and σ_2 be the simplices of K containing p and q, respectively, in their interiors. Since neither simplex is a face of the other, $\sigma = \sigma_1 \cap \sigma_2$ is a proper face of each. Let x be the image of p under the projection of σ_1 onto σ. Take p and q close enough to b that the line segment e_1 from p to x and the line segment e_2 from x to q lie in U. Now df_b is linear on the complex $L = e_1 \cup e_2$ and dg_b is an $\alpha/2$ angle-approximation to df_b on L.

We claim that the angle between the line segments $df_b(e_1)$ and $df_b(e_2)$ is at least α; it will then follow from the special case that dg_b is 1-1 on L, contradicting the assumption that $dg_b(p) = dg_b(q)$. To prove this, note that df_b is a linear isomorphism on σ_1; because projection is natural with respect to linear isomorphisms, the projection of the simplex $df_b(\sigma_1)$ onto the simplex $df_b(\sigma)$ must carry $df_b(p)$ into $df_b(x)$. This means, as noted in (2) above, that the angle between $df_b(e_1)$ and $df_b(e_2)$ is at least as large as the angle from $df_b(\sigma_1)$ to $df_b(\sigma_2)$; this angle is, in turn, at least α.

(5) Let f be an immersion; choose δ as above. Given b in $|K|$, there is a neighborhood V of b such that for any δ-approximation $g : K' \to R^n$ to f, g is 1-1 on V. The proof is as follows:

First choose an ϵ-neighborhood U of b on which df_b is an $\alpha/3$ angle-approximation to f. Then choose a neighborhood V of b, lying in U, such that for any pair σ_1, σ of simplices of K containing b such that σ is a proper face of σ_1, the projection of σ_1 onto σ carries $V \cap \sigma_1$ into U.

Now g is an $\alpha/6$ angle-approximation to f; hence g is an $\alpha/2$ angle-approximation to df_b on U. It follows that g is 1-1 on V. For suppose $g(p) = g(q)$, p and q in V. If p and q lie in the same simplex σ of K, let x be the midpoint of the line segment pq; let $e_1 = px$ and $e_2 = qx$. Since U is an ϵ-neighborhood, $U \cap \sigma$ is convex, so $L = e_1 \cup e_2$ lies in U. Now $df_b(e_1)$ and $df_b(e_2)$ make an angle of π with each other. Since $\alpha \leq \pi$, g is 1-1 on L, by the special case of our theorem, contrary to hypothesis.

The argument in the case where p and q do not lie in the same simplex of K is similar. Let p lie in $\operatorname{Int} \sigma_1$ and q in $\operatorname{Int} \sigma_2$, where $\sigma = \sigma_1 \cap \sigma_2$ is a proper face of each. Let x be the image of p under the projection of σ_1 onto σ; then x is in U. Let e_1 be the line segment px; let e_2 be the line segment qx. Then $L = e_1 \cup e_2$ lies in U, and $df_b(e_1)$ and $df_b(e_2)$ make an angle of at least α with each other. The special case of the theorem applies to show g is 1-1 on L, contrary to hypothesis.

(6) Let f be an imbedding; let δ be chosen as in (4). Applying (5), choose a finite number of compact sets C_i whose interiors cover $|K|$ such that any δ-approximation g to f is necessary 1-1 on C_i. By Exercise (a) of 3.10, there is a $\delta_0 > 0$ such that if $g : |K| \to R^n$ is a C^0 δ_0-approximation to f and g is 1-1 on each C_i, then g is a homeomorphism. Hence if g is both a δ_0 and a δ-approximation to f, g will be an imbedding.

(a) Exercise. Let $f : X \times Y \to R$ be continuous. Let A be a compact subset of Y and define $f_A(x) = \min f(x,y)$ for y in A. Prove that $f_A : X \to R$ is continuous. Conclude from this that the angle θ from σ_1 to σ_2 is a continuous function of the vertices of σ_1 and σ_2.

(b) Exercise. Derive a formula for the projection of σ_1 onto σ. Conclude that projection is continuous, and natural in the sense previously described.

(c) Exercise. Generalize the theorem to the case K infinite; δ will have now to be a positive continuous function on $|K|$, of course.

(d) Exercise. Show that given a C^r triangulation $f: K \to M$, any sufficiently close approximation g to f is also a triangulation, providing g carries Bd $|K|$ into Bd M.

§9. The Secant Map Induced by f.

Given a complex K, a finite subcomplex K_1, and a C^r map f of K into euclidean space R^n. We would like to approximate f by a C^r map g of some subdivision K' of K into R^n such that g is linear on each simplex of the corresponding subdivision K_1' of K_1.

The first step is to show that if K_1 is subdivided very carefully, the secant map g induced by f, which is automatically linear, will be a strong C^1 approximation to f on K_1 (9.1 - 9.6). The second step is to prove that g may be extended to all of $|K|$ in such a way that the extension will be a strong C^1 approximation to f (9.7 - 9.8).

9.1 Definition. Let $f : K \to R^n$ be a C^r map. Let s be any simplex which is contained in a simplex of K. The linear map $L_s: s \to R^n$ which equals f on the vertices of f is called the *secant map induced by* f.

Similarly, if K' is a subdivision of K, then the *secant map induced by* f, $L_{K'}: K' \to R^n$, is the linear map which equals f on the vertices of K'. Note that if f carries a simplex into H^n or R^{n-1}, so does the secant map induced by f.

9.2 Definition. The *radius* $r(\sigma)$ of the simplex σ is the minimum distance from the barycenter $\tilde{\sigma}$ to Bd σ. The *diameter* $d(\sigma)$ of σ is the length of the longest edge. The *thickness* $t(\sigma)$ of σ is $r(\sigma)/d(\sigma)$.

9.3 Lemma. Let $f : \sigma \to R^n$ be a C^r map; let δ and t_0 be positive numbers. There is an $\epsilon > 0$ such that for any simplex s contained in σ having diameter less than ϵ and thickness greater than t_0, the secant map L_s is a δ-approximation to $f|s$.

Proof. Given b in σ, let $F_b(x) = f(b) + Df(b) \cdot (x - b)$. Then $f(x) - F_b(x) = h(x, b)$, where $\|h(x, b)\|/\|x - b\| \to 0$ uniformly as $\|x - b\| \to 0$. (See (2) of 1.10.) Also, $Df(x) - DF_b(x) = Df(x) - Df(b)$. It follows that there is an ϵ_1 independent of b such that F_b is a $\delta/2$ approximation to f on the ϵ_1 neighborhood of b. We require ϵ to be less than ϵ_1; this implies that $\|h(x, b)\| < \delta/2$ for $\|x - b\| < \epsilon$. We further require that $\|h(x, b)\|/\|x - b\| < \delta t_0/4$ for $\|x - b\| < \epsilon$.

We prove that if s has thickness at least t_0 and diameter less than ϵ, then L_s is a $\delta/2$ approximation to $F_b|s$, where b is the barycenter of s. This suffices to prove our theorem, since $F_b|s$ is a $\delta/2$ approximation to $f|s$.

(1) We need a preliminary result: If L and F are linear maps of a simplex s into R^n and $\| L(x) - F(x)\| < \beta$ for all x, then

$$\| DL(x) \cdot \vec{u} - DF(x) \cdot \vec{u} \| < 2\beta/r(s)$$

for all unit vectors $\vec{u}$ in the plane of σ.

For let b be the barycenter of s. Since L is linear, $L(x) = L(b) + DL(b) \cdot (x-b)$, and $DL(x) = DL(b)$. Similarly for DF. Then

$$[DL(x) - DF(x)] \cdot \vec{u} = [DL(b) - DF(b)] \cdot (x-b)/\| x - b \|$$

for some x in Bd s. This equals

$$\{[L(x) - L(b)] - [F(x) - F(b)]\}/\| x - b \| ,$$

so

$$\| DL(x) \cdot \vec{u} - DF(x) \cdot \vec{u} \| < 2\beta/\| x - b \| \leq 2\beta/r(s)$$

(2) Proof of the lemma. Let $v_0, \ldots, v_p$ be the vertices of s; let $x = \Sigma\, \alpha_i v_i$ be the general point of s. Since L_s and F_b are linear,

$$L_s(x) = \Sigma\, \alpha_i\, L_s(v_i) = \Sigma\, \alpha_i\, f(v_i)$$

and

$$F_b(x) = \Sigma\, \alpha_i\, F_b(v_i) \quad ,$$

so that

$$\| L_s(x) - F_b(x) \| = \| \Sigma\, \alpha_i\, h(v_i, b) \| \leq \max \|h(v_i, b)\|.$$

Now $\|h(x,b)\| < \delta/2$ whenever $\|x - b\| < \epsilon$, so that $\| L_s(x) - F_b(x) \| < \delta/2$, which is the first part of our desired conclusion. Furthermore, $\| h(x,b)\|/\| x - b \| < \delta t_0/4$ whenever $\| x - b \| < \epsilon$, so that

$$\| L_s(x) - F_b(x) \| < \max \|v_i - b\| \delta t_0/4 \leq d(s)\ \delta t_0/4 \leq \delta\ r(s)/4 ,$$

since $t_0 \leq r(s)/d(s)$. From the preliminary result (1) we conclude that L_s is a $\delta/2$ approximation to $F_b|s$.

(a) Exercise*. Show by means of an example that the theorem fails without the hypothesis that the thickness of s is bounded away from zero.

9.4 Lemma. Let K be a finite complex. There is a $t_0 > 0$ such that K has arbitrarily fine subdivisions for which the minimal simplex thickness is at least t_0.

Proof. We first remark that if the theorem is true for the complex K, it is true for any complex which is isomorphic to K. For let $f : K \to K_1$ be a linear isomorphism. For each simplex of K, the ratio $\| f(x) - f(y) \| / \| x-y \|$ is bounded, and bounded above zero. Choose α and β so that

$$0 < \alpha < \| f(x) - f(y) \| / \| x-y \| < \beta$$

whenever x and y belong to the same simplex of K. Given $\epsilon > 0$, let K' be a subdivision of K of diameter less than ϵ and thickness greater than t_0 (i.e., the maximal diameter of a simplex of K' is less than ϵ, and the minimal thickness of such a simplex is greater than t_0). The corresponding subdivision of K_1 has diameter less than $\beta\epsilon$ and thickness greater than $\alpha t_0/\beta$, as the reader may verify, so our result follows.

Now the complex K is isomorphic to a subcomplex of some "standard simplex" — the "standard simplex" of dimension p being the one having $\epsilon_0, \epsilon_1, \ldots, \epsilon_p$ as vertices, where ϵ_0 is the origin and the others are the natural basis vectors for R^p. Hence it will suffice to prove the theorem in the case where K is this standard simplex.

Let J be the cell complex whose polytope is R^p, whose cells are obtained as follows: If $i_0, i_1, \ldots, i_p$ are integers, consider the cube in R^p defined by the equation

$$C(i_1, \ldots, i_p) = \{x \mid i_j \leq x^j \leq i_j + 1 \quad \text{for} \quad j = 1, \ldots, p\} ,$$

and intersect it with the region

$$R(i_0) = \{x \mid i_0 \leq x^1 + \ldots + x^p \leq i_0 + 1\} .$$

The resulting set will be a cell. (To illustrate, the accompanying illustration shows $C(0, 0, 0)$ intersected with $R(0)$, $R(1)$, and $R(2)$.) Take J as the collection of all such cells and their faces. The cell complex

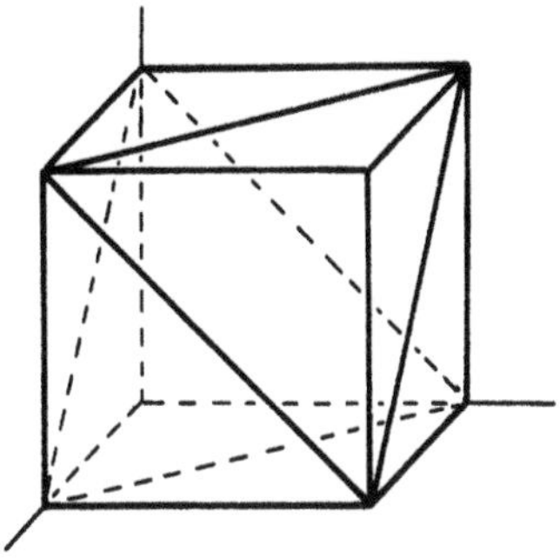

J has two properties which are important for us:

(1) Any cell of J is the image of one of the cells contained in the unit cube $C(0,\dots,0)$, under a translation of R^p.

(2) The simplex Σ_m spanned by $0, m\epsilon_1, \dots, m\epsilon_p$ is the polytope of a subcomplex of J, for each positive integer m.

Now we subdivide J into a simplicial complex L by the step-by-step process described in Lemma 7.8; at each step we use the centroid $\tilde{c}$ of c as the interior point of the cell c which determines the subdivision. It follows that conditions (1) and (2) hold for the complex L as well. As a result, the simplices of L have a minimal thickness $t_0 > 0$, and maximal diameter d.

The similarity transformation on R^p which carries x into x/m does not change the thickness of any simplex, and it multiplies the diameter by $1/m$. Therefore the image of Σ_m under this transformation will be a rectilinear triangulation of Σ_1 of thickness at least t_0 and diameter at most d/m. Since m is arbitrary, the theorem is proved.

(a) Exercise*. Let K consist of a 2-simplex and its faces. Let K_1 be the barycentric subdivision of K, K_2 the barycentric subdivision of K_1, etc. Show that the thickness of K_n approaches zero as $n \to \infty$.

9.5 Unsolved problem*. Generalize this theorem to non-finite complexes K. One would expect t_0 and ϵ to be continuous functions on $|K|$, but it is not entirely clear even what the proper conjecture should be.

9.6 Theorem. Let $f : K \to R^n$ be a C^r map; K finite. Given $\delta > 0$, there is a subdivision K' of K such that the secant map $L_{K'} : K' \to R^n$ induced by f is a δ-approximation to f.

We now obtain the following generalization of this theorem, which will follow immediately from the lemma which succeeds it:

9.7 Theorem. Let $f : K \to R^n$ be a C^r map; let K_1 be a finite subcomplex of K. Given $\delta > 0$, there is a δ-approximation $g : K' \to R^n$ to f such that

(1) g equals the secant map induced by f on K_1'.

(2) g equals f outside $St(K_1,K)$.

(3) K' equals K outside $St(K_1,K)$.

Here K_1' is the subdivision of K_1 induced by K'; and condition (3) means that every simplex of K outside $St(K_1,K)$ appears in K'.

9.8 Lemma. Let $f : K \to R^n$ be a C^r map. Let K_1 be a finite subcomplex of K. Given $\epsilon > 0$, there is a $\delta > 0$ such that any δ-approximation $g : K_1' \to R^n$ to $f|K_1$ can be extended to an ϵ-approximation $h : K' \to R^n$ to f. Here K' is the standard extension of K_1' (see 7.12), and h equals f outside $St(K_1,K)$.

Proof. The subdivision K' of K is specified, and the map h is defined on $|K_1|$ and on $|K| - St(K_1,K)$. The only problem remaining is to extend h to each simplex of K' whose interior lies in $St(K_1,K) - |K_1|$, and to see how good an approximation h is. We carry out this extension step-by-step, first to the 1-simplices, then the 2-simplices, and so on. For this reason, it will suffice to prove our lemma for the special case of a single simplex:

Special case. Let L be a complex consisting of a simplex σ and its faces; let $L_1 = Bd\ \sigma$. Given $\epsilon > 0$, there is $\delta > 0$ such that any δ-approximation $g: L_1' \to R^n$ to $f|L_1$ may be extended to an ϵ-approximation. $h: L' \to R^n$ to $f|L$, where L' is the standard extension of L_1'

<u>Proof of special case</u>. An obvious extension of g would be the one mapping the line segment $y\tilde{\sigma}$ linearly onto the line segment $g(y)\ f(\tilde{\sigma})$, where $\tilde{\sigma}$ is the barycenter of σ. But there are difficulties here, for the resulting map will not be differentiable at $\tilde{\sigma}$. We modify this definition, "tapering off" between $g(\text{Bd}\ \sigma)$ and $f(\tilde{\sigma})$ by a smooth function $\alpha(t)$, rather than linearly.

Let $\alpha(t)$, as usual, be a monotonic C^∞ function which equals 0 for $1/3 \leq t$, and equals 1 for $t \geq 2/3$. If x is in σ, then $x = ty + (1-t)\,\tilde{\sigma}$, where y is in $\text{Bd}\ \sigma$ and $0 \leq t \leq 1$; t is uniquely determined, and y is unique if $x \neq \tilde{\sigma}$. Furthermore, if L_1' is any subdivision

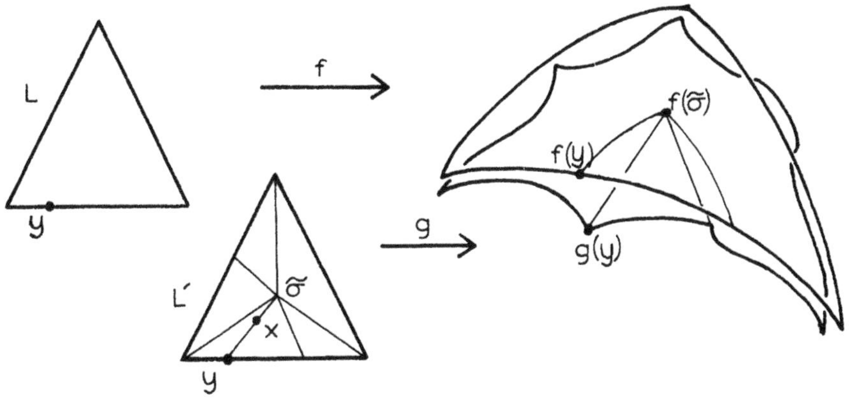

of $\text{Bd}\ \sigma$, and L' is the standard extension, then y and t are C^∞ functions of x on each simplex of L', except at the point $x = \tilde{\sigma}$.

Let g be a C^r map of a subdivision L_1' of L_1 into R^n. Define

$$h(x) = f(x) + \alpha(t(x)) \cdot [g(y(x)) - f(y(x))] \quad .$$

Since $\alpha(t(x)) = 0$ for x in a neighborhood of $\tilde{\sigma}$, h is of class C^r on each simplex of L'. We show that h is automatically a good approximation to f if g is a good approximation to $f|\text{Bd}\ \sigma$.

First note that $\| h(x) - f(x) \| \leq \| g(y) - f(y) \|$, so h will be as good a C^0 approximation to f as g is.

Second, we compute $\partial(h-f)/\partial x$ for $t \geq 1/3$. It equals

$$\alpha'(t)[g(y) - f(y)] \cdot \partial t/\partial x \;+\; \alpha(t)[\partial g/\partial y - \partial f/\partial y] \cdot \partial y/\partial x \;,$$

where g and f are written as column vectors, as usual. Now $\alpha'(t)\cdot\partial t/\partial x$ and $\alpha(t)\cdot\partial y/\partial x$ are bounded independent of the choice of the subdivision L_1' and the map g. Hence we may make $\partial(h-f)/\partial x$ as small as desired, simply by requiring $|g(y) - f(y)|$ and $|\partial g/\partial y - \partial f/\partial y|$ to be sufficiently small.

(a) Exercise. Compute exactly how small $|g(y) - f(y)|$ and $|\partial g/\partial y - \partial f/\partial y|$ need to be in order that $|\partial(h - f)/\partial x|$ should be less than δ.

(b) Exercise. Complete the proof of the lemma using the result for the special case proved above.

(c) Exercise. Show that the following addendum to the lemma holds: Suppose that whenever f carries a simplex σ_1 of K_1 into the plane $\mathcal{P}$, then g carries σ_1 into $\mathcal{P}$, as well. Prove that whenever f carries a simplex σ of K into $\mathcal{P}$, then h carries σ into $\mathcal{P}$, as well.

(d) Exercise. Let $\mathcal{P}_1, \ldots, \mathcal{P}_k$ be a finite number of the planes in R^n. Prove Theorem 9.7, with the additional conclusion that each simplex σ of K which f carries into one of the planes $\mathcal{P}_i$ is also carried by g into $\mathcal{P}_i$.

(e) Exercise. Show that (c) does not hold if you replace the plane $\mathcal{P}$ throughout by a half-plane, for instance H^m.

§10. Fitting Together Imbedded Complexes.

Suppose now we have C^r imbeddings of two complexes into the differentiable manifold M, whose images in M overlap. We would like to prove that by altering these imbeddings slightly, we may make their images fit together nicely, i.e., intersect in a subcomplex, after suitable subdivisions.

If M is euclidean space, this is not too difficult; the construction is carried out in 10.2. The basic idea is to alter the imbeddings so that they are *linear* near the overlap (using the results of §9), for we know from §7 that two rectilinear complexes always intersect nicely.

If M is not euclidean space, we must use the coordinate systems on M, which look like euclidean space, and carry out this alteration step-by-step. The process is not basically more difficult, but it is somewhat more complicated (10.3 - 10.4). After this, however, the two main theorems of Chapter II fall out almost trivially (10.5 - 10.6).

10.1 Definition. Let $f_1: K_1 \to R^n$ and $f_2: K_2 \to R^n$ be homeomorphisms; let $f_1(|K_1|)$ and $f_2(|K_2|)$ be closed subsets of their union. (K_1, f_1) and (K_2, f_2) are said to *intersect in a subcomplex* if the inverse images of $f_1(|K_1|) \cap f_2(|K_2|)$ are polytopes of subcomplexes L_1 and L_2 of K_1 and K_2, respectively, and $f_2^{-1}f_1$ is a linear isomorphism of L_1 with L_2. They intersect in a *full subcomplex* if L_i contains every simplex of K_i whose vertices belong to L_i, either for $i = 1$ or for $i = 2$. In such a case, there is a complex K and a homeomorphism $f : K \to R^n$ such that the following diagram is commutative:

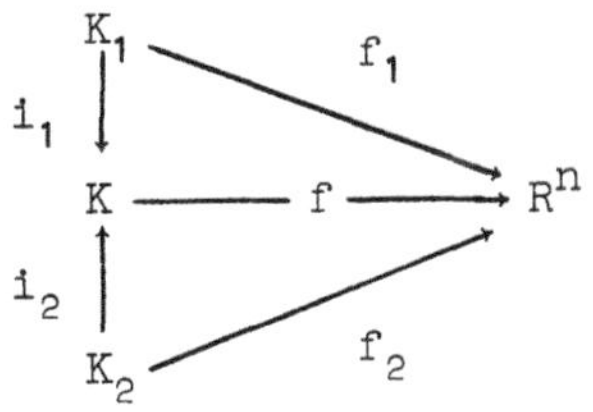

Here i_1 and i_2 are linear isomorphisms of K_1 and K_2 with subcomplexes of K whose union equals K. The pair (K,f) is unique up to a linear isomorphism; it is called the union of (K_1,f_1) and (K_2,f_2). Let L denote the subcomplex $i_1(K_1) \cap i_2(K_2)$.

We will be interested in this situation in the case where $f_1: K_1 \to R^n$ and $f_2: K_2 \to R^n$ are, in addition, C^r imbeddings. The map $f: K \to R^n$ will be a C^r non-degenerate homeomorphism in this case, but it may not be an imbedding, because the immersion condition may fail. That is, for some point b of $|L|$, the map df_b may not be 1-1, even though f is. (See the example of 8.1.) However, if f_j should happen to be linear on $\overline{St}(L_j,K_j)$ for $j = 1$ and 2, then the union is necessarily an imbedding, for in this case $df_b = f|\overline{St}\ b$ for b in $|L|$; and f is given to be 1-1. This fact will be of importance in what follows. Another case in which the union is an imbedding is given in Exercise (c).

(a) Exercise. Let $f_j: K_j \to R^n$ be a homeomorphism, $j = 1,2$. If $f_1: K_1 \to R^n$ and $f_2: K_2 \to R^n$ intersect in a subcomplex, prove that $f_1: K_1' \to R^n$ and $f_2: K_2' \to R^n$ intersect in a full subcomplex, where K_1' and K_2' are the barycentric subdivisions of K_1 and K_2, respectively.

(b) Exercise. Prove existence and essential uniqueness of the union (K,f).

Outline: Form an abstract complex K whose vertices are the points $f_j(v^j)$, where v^j is any vertex of K_j $(j = 1,2)$. If $v_0^j \ldots v_p^j$ is a simplex of K_j, let the collection $\{f_j(v_0^j),\ldots,f_j(v_p^j)\}$ be a simplex of K. Then take a geometric realization of this abstract complex, and define f. To show f is a homeomorphism, you will need to consider the limit set of f.

(c) Exercise. Let $f_j: K_j \to R^n$ be C^r imbeddings $(j = 1,2)$ whose intersection is a full subcomplex. Let $f: K \to R^n$ be their union, and let i_j be the inclusion of K_j in K. If $|K|$ is the union of $\text{Int}\ i_1(|K_1|)$ and $\text{Int}\ i_2(|K_2|)$, then $f: K \to R^n$ is a C^r imbedding.

10.2 Lemma. Let $f: P \to R^m$ be a C^r map. Let Q and A be subcomplexes of P whose union is P; let $f|Q$ and $f|A$ be C^r imbeddings whose images are closed in R^m; let A be finite. Given

$\delta > 0$, there is a δ-approximation $g : P' \to R^m$ to f such that $g|Q'$ and $g|A'$ intersect in a full subcomplex, and their union is a C^r imbedding.

Furthermore, g equals f, and P' equals P, outside $St^3(f^{-1}f(A),P) = St(\overline{St}(\overline{St}(f^{-1}f(A))))$, all stars being taken in P.

Proof. Let P_0 be the complex whose polytope is $\overline{St}(f^{-1}f(A))$; it consists of all simplices whose images intersect $f(A)$, along with their faces, so it contains A. We assume δ is less than half the distance from $f(P) - f(P_0)$ to $f(A)$, so that any δ-approximation g to f cannot carry a point of $P - P_0$ into a point of $g(A)$. We also assume that δ is small enough that for any δ-approximation $g : P' \to R^m$ to f, $g|Q'$ and $g|A'$ are C^r imbeddings.

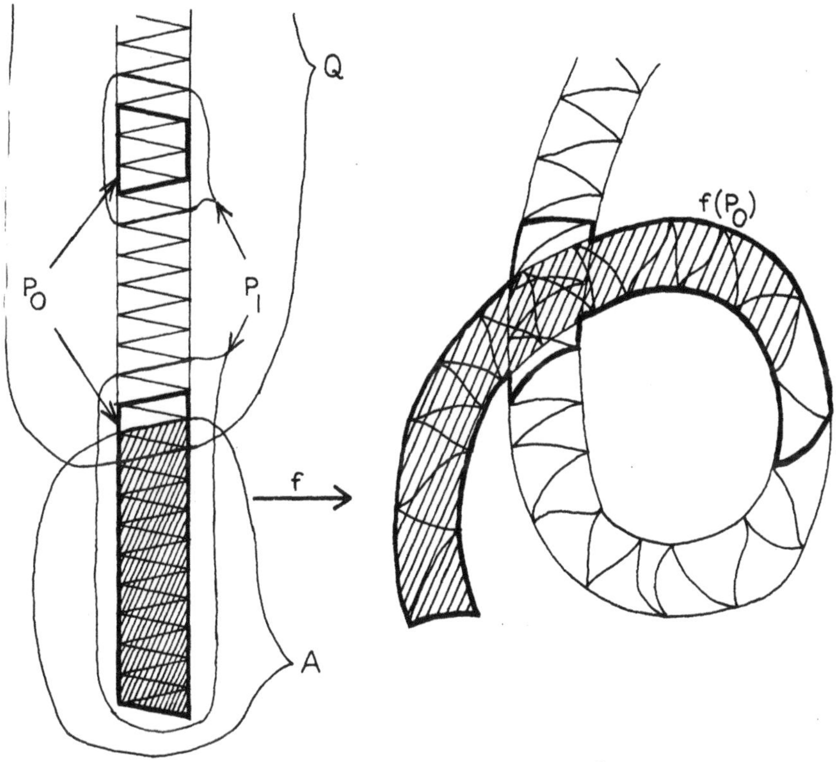

Let P_1 be the subcomplex of P whose polytope is $\overline{St}(P_0,P)$. By Theorem 9.7, there is a δ-approximation $g : P' \to R^m$ to f which equals the secant map induced by f on the subcomplex P_1'; further, g equals f and P' equals P outside $St(P_1,P)$.

Let Q_1 be the part of P_1 that lies in Q, so that $P_1 = Q_1 \cup A$. Now $g|Q_1'$ and $g|A'$ are linear imbeddings, so the images are finite rectilinear complexes in euclidean space. By Theorem 7.10, one may choose subdivisions of these complexes whose union is a complex. These subdivisions induce, via g^{-1}, a subdivision P_1'' of P_1' such that $g|Q_1''$ and $g|A''$ intersect in a subcomplex. We take the barycentric subdivision P_1''' of P_1'', so that $g|Q_1'''$ and $g|A'''$ intersect in a full subcomplex (see Exercise (a) of 10.1). Extend this subdivision of P_1' in the standard way to a subdivision P''' of P' (see 7.12).

We now claim that $g|Q'''$ and $g|A'''$ intersect in a subcomplex. For suppose $g(\sigma_1)$ intersects $g(\sigma_2)$, where σ_1 is in Q''' and σ_2 is in A'''. Then σ_1 must lie in $|P_0|$, by our original assumption on δ, so that in particular, σ_1 lies in $|Q_1|$. But $g|Q_1'''$ and $g|A'''$ intersect in a subcomplex. Hence $g|Q'''$ and $g|A'''$ intersect in a subcomplex. A similar remark shows this subcomplex is full.

It follows that the union of $g|Q'''$ and $g|A'''$ is a C^r imbedding; we apply the remark at the end of 10.1. For if $g(b_1) = g(b_2)$, for b_1 in $|Q|$ and b_2 in $|A|$, then b_1 must lie in $|P_0|$, by choice of δ. But g is <u>linear</u> on P_1''', which contains $\overline{St}(|P_0|, P''')$. Hence the union of $g|Q'''$ and $g|A'''$ is a C^r imbedding.

<u>10.3 Corollary</u>. Let M be a non-bounded C^r submanifold of some euclidean space. The preceding lemma holds if R^m is replaced by M throughout, providing we assume that $f(A)$ is contained in some coordinate system (U,h) on M.

<u>Proof</u>. First take the collection of those simplices of P which intersect $f^{-1}f(A)$ and subdivide them finely enough that each has diameter less than one-fourth the distance from $f^{-1}f(A)$ to $f^{-1}(M - U)$. Extend to a subdivision P' of P in the standard way. The result is that f carries $\overline{St}^4(f^{-1}f(A), P')$ into U.

We then take the subcomplex J of P' whose polytope is $\overline{St}^4(f^{-1}f(A), P')$ and consider the C^r map $hf : J \to R^m$. The preceding lemma applies. There is an approximation $G : J' \to R^m$ to hf; G equals

hf and J' equals J outside $St^3(f^{-1}f(A),P')$. Hence we may define $g = h^{-1}G$ on $St^4(f^{-1}f(A),P')$, and $g = f$ outside $\overline{St}^3(f^{-1}f(A),P')$ to obtain the desired map; no simplex of P' outside J needs to be subdivided at all.

The preceding lemma guarantees that G is a homeomorphism. To be sure g is, we need to have g approximate f closely enough that $g(\overline{St}^3(f^{-1}f(A),P'))$ and $g(P - St^4(f^{-1}f(A),P'))$ are disjoint. Similarly, the lemma guarantees that G is a C^r imbedding; the fact that $h^{-1}G$ is an imbedding follows from Exercise (a) of 8.3. Finally, one uses Exercise (c) of 10.1 to prove that g is an imbedding.

(a) Exercise. Note that the following additional conclusion holds in the preceding corollary: $g(x) = f(x)$ if $f(x)$ lies outside the coordinate neighborhood (U,h) chosen containing $f(A)$.

10.4 Theorem. Let M be a non-bounded C^r submanifold of R^n. Let $f : K \to M$ and $g : L \to M$ be C^r imbeddings whose images are closed in M. Given $\delta(x) > 0$, there are δ-approximations $f': K' \to M$ and $g': L' \to M$ to f and g respectively, which intersect in a full subcomplex, such that their union is a C^r imbedding. (δ is to be continuous on the disjoint union of $|K|$ and $|L|$.)

Proof. Assume g carries each simplex of L into a coordinate neighborhood on M. (See Exercise (a) of 7.13.) Order the simplices of $L : A_1, A_2, \ldots$ in such a way that each simplex is preceded in the ordering by all its faces. Choose a neighborhood V_i of $g(A_i)$ in M, for each i, so that $\{\overline{V}_i\}$ is a locally-finite collection of compact subsets of M.

(*) Assume δ is small enough that for any δ-approximations f' and g' to f and g, respectively, $g'(A_i)$ is contained in V_i and is disjoint from $f'f^{-1}(M - V_i)$.

Let $f_0 = f$ and $g_0 = g$.

Induction hypothesis. Suppose $f_i: K_i \to M$ and $g_i: L_i \to M$ are C^r imbeddings, where K_i is a subdivision of K and L_i is a subdivision of L; and f_i and g_i are $(1 - 1/2^i) \cdot \delta(x)$ approximations to f and g, respectively. Furthermore, if J_i denotes the subcomplex of L_i whose polytope is $A_1 \cup \ldots \cup A_i$, suppose that $f_i: K_i \to M$ and $g_i | J_i$ intersect in a full subcomplex, and their union is an imbedding.

Let $h : Q \to M$ be the C^r imbedding which is the union of (K_i, f_i) and $g_i | J_i$. Consider K_i and J_i as subcomplexes of Q, so that $Q = K_i \cup J_i$. Now A_{i+1} is the polytope of a subcomplex of L_i, which we also denote by A_{i+1}; let P be the complex obtained by identifying each point of Bd A_{i+1} with the corresponding point of J_i. We may have to subdivide A_{i+1} in order that P should be a complex; we assume this done without change of notation. Extend h to P by letting it equal g_i on A_{i+1}.

Now apply the preceding corollary. The map $h : P \to M$ is a C^r map; $h|Q$ and $h|A_{i+1}$ are C^r imbeddings, and $h(A_{i+1})$ is contained in a coordinate system on M. Given $\epsilon > 0$, there is an ϵ-approximation $h': P' \to M$ to h such that $h'|Q'$ and $h'|A'_{i+1}$ are C^r imbeddings which intersect in a full subcomplex such that their union is an imbedding. This union is the same as the union of $h'|K'_i$ and $h'|J'_i \cup A'_{i+1}$, so the union of the latter pair is also an imbedding.

If ϵ is small enough, $h': K'_i \to M$ will be a $\delta(x)/2^{i+1}$ approximation to h, so that $f_{i+1} = h'|K'_i$ is an $(1 - 1/2^{i+1})\delta(x)$ approximation to f, as desired.

Further, if ϵ is small enough, $h': J'_i \cup A'_{i+1} \to M$ may be extended to a C^r map $g_{i+1}: L_{i+1} \to M$ which is a $\delta(x)/2^{i+1}$ approximation to $g_i: L_i \to M$. Here we use Lemma 9.8 again. Thus the induction hypothesis is satisfied, so that f_i and g_i are defined for all i.

We would like to define $f' = \lim_{i \to \infty} f_i$ and $K' = \lim K_i$; and similarly for g' and L'. The induction hypothesis does not contain enough information to enable us to do this, so let us examine the definitions of f_{i+1} and g_{i+1} more closely.

Corollary 10.3 tells us that Q' equals Q and h' equals h, outside $\mathrm{St}^3(h^{-1}h(A_{i+1}), P)$. Hence f_{i+1} equals f_i, and K_{i+1} equals K_i, for simplices outside $\mathrm{St}^3(f_i^{-1}\, g_i(A_{i+1}), K_i)$. Now by the original assumption $(*)$ on δ, f_i carries a point x of K into $g_i(A_{i+1})$ only if f carried x into V_i. Thus it is certain that f_{i+1} equals f_i, and K_{i+1} equals K_i, outside $\mathrm{St}^3(f^{-1}(V_i), K)$. These sets are a locally-finite collection of subsets of K, so that $\lim f_i$ and $\lim K_i$ do exist (see 7.13).

Similarly, g_{i+1} equals g_i, and L_{i+1} equals L_i, outside $St^3(g_i^{-1}\, g_i(A_{i+1}), L_i) = St^3(A_{i+1}, L_i) \subset St^3(A_{i+1}, L)$. Hence the limits exist in this case as well. The limiting maps will automatically be $\delta(x)$-approximations to f and g, respectively. That their union is a C^r imbedding is easily checked.

(a) Exercise. By examining the proof of this theorem more closely, draw the following additional conclusion: If V is any neighborhood of $g(|L|)$, then we may so choose K' and f' that K' equals K, and f' equals f, outside $St^3(f^{-1}(V), K)$.

(b) Exercise. Apply Exercise (a) of 10.3 and Exercise (d) of 9.8 to strengthen 10.4 as follows: Assume f and g carry subcomplexes of K and L, respectively, into the non-bounded C^r submanifold N of M. Then f' and g' may be chosen so as to carry these subcomplexes into N.

10.5 Theorem. Let M be a C^r manifold; let $f : K \to M$ and $g : L \to M$ be C^r triangulations of M. There are subdivisions of K and L which are linearly isomorphic.

In fact, given $\delta(x) > 0$, there are δ-approximations $f': K' \to M$ and $g': L' \to M$ which are C^r triangulations of M, such that $(f')^{-1} g'$ is a linear isomorphism of L' with K'.

Proof. In the non-bounded case, we choose $\delta(x) > 0$ so that any δ-approximations to f and g are necessarily onto maps, using Lemma 3.11. Our result then follows from Theorem 10.4. In the other case, we consider M as a submanifold of $D(M)$, and apply Exercise (b) of 10.4 to assure that $f' : K' \to D(M)$ carries $\text{Bd}\,|K|$ into $\text{Bd}\, M$. If δ is small enough, $f'(\text{Bd}\,|K|) = \text{Bd}\, M$, so that $f'(|K|) \subset M$. Again, if δ is small enough, $f'(|K|) = M$. Similarly, if δ is small enough, $g'(|L|) = M$.

10.6 Theorem. If M is a non-bounded C^r manifold, M has a C^r triangulation. If M is a manifold having a boundary, any C^r triangulation of the boundary may be extended to a C^r triangulation of M. (If

$f : J \to \text{Bd}\, M$ is a C^r triangulation of Bd M, an extension of f is a C^r triangulation $g : L \to M$ of M such that $g^{-1} f$ is a linear isomorphism of J with a subcomplex of L.)

Proof. Let M be non-bounded. Cover M by $m + 1$ coordinate systems (U_i, h_i), $i = 0,\ldots,m$, chosen so that $h_i(U_i)$ is the union of a countable collection $h_i(V_{ij})$ of disjoint bounded open sets in R^m (see 2.8). Let C_{ij} be a closed set contained in V_{ij} such that the collection $\{C_{ij}\}$ covers M. Let K_{ij} be a finite rectilinear complex in R^m contained in $h_i(V_{ij})$, and containing a neighborhood of $h_i(C_{ij})$; let K_i be the complex $\cup_{j=1}^{\infty} K_{ij}$. Then $h_i^{-1}: K_i \to M$ is a C^r imbedding whose image contains a neighborhood of $\cup_j C_{ij}$. Let $g_i: K_i' \to M$ be a $\delta_i(x)$-approximation to h_i^{-1}, for $i = 0,\ldots,m$, chosen so that $(K_0', g_0),\ldots,(K_m', g_m)$ intersect in a full subcomplex and their union is an imbedding. This union will be a C^r triangulation of M if $\delta_i(x) > 0$ is chosen small enough that $g_i(|K_i|)$ contains $\cup_j C_{ij}$ (see Exercise (b) of 3.11).

Now suppose M has a boundary; let $f : J \to \text{Bd}\, M$ be a C^r triangulation. We triangulate the space $|J| \times [0,1)$ as follows: Suppose J lies in R^n. For each simplex σ of J, and each positive integer n, consider the cells $\sigma \times [1 - 1/n, 1 - 1/n{+}1]$ and $\sigma \times (1 - 1/n)$, which are contained in R^{n+1}. The collection of all such cells is a cell complex whose polytope is $|J| \times [0,1)$; this cell complex may be subdivided into a simplicial complex without subdividing any cell $\sigma \times (1 - 1/n)$ of the second kind. We denote the resulting simplicial complex by K.

Let K_0 be the subcomplex of K whose polytope is $|J| \times [0,5/6]$; let $P : \text{Bd}\, M \times [0,1) \to M$ be a product neighborhood of the boundary. The composite

$$J \times [0,1) \xrightarrow{f \times i} \text{Bd}\, M \times [0,1) \xrightarrow{P} M \quad ,$$

when restricted to K_0, is a C^r imbedding $g : K_0 \to M$ whose image is closed in M and contains $P(\text{Bd}\, M \times [0,4/5])$ in its interior.

Let $h : L \to M$ be a C^r triangulation of the non-bounded manifold Int M. By subdividing L if necessary, we may assume that any simplex of L which intersects the set $P(\text{Bd}\, M \times (4/5))$ is disjoint from the set $P(\text{Bd}\, M \times [0,3/4])$, using Exercise (a) of 7.13. Let L_0 be the subcomplex of L consisting of simplices whose images under h intersect

$M - P(\text{Bd } M \times [0,4/5))$, and their faces. Then $h : L_0 \to M$ is a C^r imbedding whose image is closed in M, contains $M - P(\text{Bd } M \times [0,4/5])$ in its interior, and is contained in $M - P(\text{Bd } M \times [0,3/4])$. By Theorem 10.4, there are

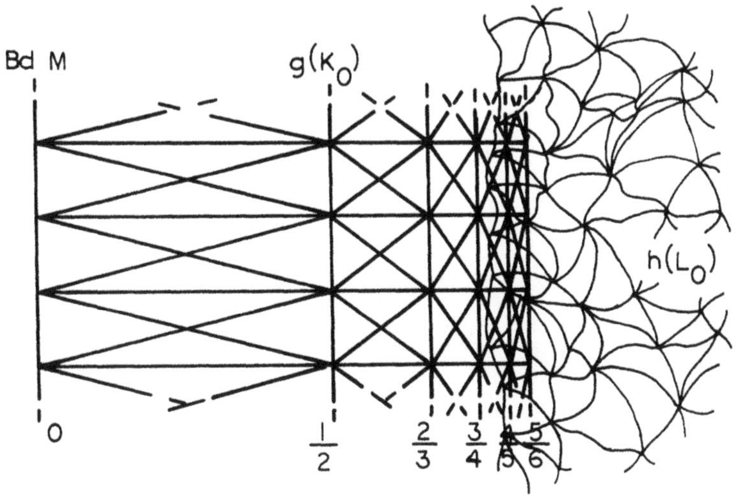

approximations $g' : K_0' \to M$ and $h' : L_0' \to M$ to g and h, respectively, whose intersection is a full subcomplex such that the union is an imbedding. According to Exercise (a) of 10.4, we may assume g' equals g, and K_0' equals K_0, on $|J| \times [0,1/2]$. The union of (K_0',g') and (L_0',h') will thus be the required C^r triangulation of M, providing $g'(|K_0|)$ and $h'(|L_0|)$ cover M.

Now $g(|K_0|)$ contains $P(\text{Bd } M \times [1/2,4/5])$ in its interior, and $h(|L_0|)$ contains $M - P(\text{Bd } M \times [0,3/4])$ in its interior. Exercise (b) of 3.11 applies to show that the images of g' and h' will contain these two closed sets, if the approximation is good enough; and $g'(|K_0|)$ automatically contains $P(\text{Bd } M \times [0,1/2])$. Thus in this case $g'(|K_0|)$ and $h'(|L_0|)$ do cover M.

10.7 Problem*. Let A be a closed subset of the manifold M. Let $f: K \to M$ be a C^r imbedding such that $f(|K|)$ contains A in its interior. If K_0 is a subcomplex of K such that $f|K_0$ triangulates A, then $f|K_0$ may be extended to a triangulation of M.

10.8 Problem*. Let M be a non-bounded C^r submanifold of R^n. Let $r\colon W \to M$ be the retraction of a neighborhood W of M in R^n onto M which was defined in 5.5. Prove there is a rectilinear complex K in R^n lying in W such that $r|K$ is a C^r triangulation of M.

Hint: If M is compact, one may take the secant map induced by a suitably chosen smooth triangulation $f\colon K \to M$. If M is not compact, this will not suffice. Instead write K as the union of finite subcomplexes K_i with $|K_i| \subset \operatorname{Int} |K_{i+1}|$, and use the fact that if $g\colon K' \to R^n$ is linear on K_i', then the secant map induced by g equals g on K_i'.

10.9 Problem*. Let $f\colon K \to M$ be a C^r map. Prove that any map sufficiently close to f is isotopic to f. Specifically, prove the following: Given $\epsilon(x) > 0$, there exists $\delta(x) > 0$ such that for any δ-approximation $g\colon K' \to M$ to f, there is a map $f_t\colon |K| \times I \to M$ such that

(1) f_t is of class C^r on the cell $\sigma \times I$, for each simplex σ of K',

(2) $f_t\colon K' \to M$ is an ϵ-approximation to f, for each t, and

(3) $f_0 = f$ and $f_1 = g$.

Hint: Use the techniques of Problem 5.15.

10.10 Definition. Let $N_1, \ldots, N_k$ be non-bounded submanifolds of the non-bounded n-manifold M. They are said to intersect transversally if for each x in the intersection of any collection $N_{i_1}, \ldots, N_{i_j}$ of these submanifolds, there is a coordinate system $h\colon U \to R^n$ about x in M such that $h(U \cap N_{i_k})$ is contained in a plane in R^n whose dimension is that of N_{i_k}, for each k.

10.11 Problem. Generalize Theorem 10.4 as follows: (1) If H is a subcomplex of L and $f^{-1}g$ is a linear isomorphism of H onto a subcomplex of K, then we may require that $(f')^{-1}g' = f^{-1}g$ on $|H|$.

(2) Let $N_1, \ldots, N_p$ be non-bounded submanifolds of M which intersect transversally. If f and g carry subcomplexes $K_{(i)}$ and $L_{(i)}$ of K and L, respectively, into the submanifolds N_i, then we may require f' and g' to carry these subcomplexes into N_i.

Hint: Condition (2) will follow readily from Exercise (d) of 9.8. To obtain condition (1), proceed as follows: Assume H is a full subcomplex. Let $A_1, A_2, \ldots$ denote the simplices of L not in H. Alter the induction hypothesis in the proof of 10.4 as follows: Suppose $f_i: K_i \to M$ and $g_i: L_i \to M$ are C^r imbeddings which are $(1 - 1/2^i)\ \delta(x)$ approximations to f and g, respectively, and $f_i^- g_i = f^{-1} g$ on $|H|$. Let J_i denote the subcomplex of L_i whose polytope is $H \cup A_1 \cup \ldots \cup A_i$; suppose that $f_i: K_i \to M$ and $g_i|J_i$ intersect in a full subcomplex and their union is a C^r imbedding.

10.12 Theorem. Let $f: K \to M$ be a smooth triangulation of the non-bounded manifold M. Let $g: L \to M$ be a C^r imbedding, with $g(L)$ closed in M. Given $\delta(x) > 0$, there is a δ-approximation $g': L' \to M$ to g such that $f^{-1}g': L' \to K$ is linear.

(1) If $f^{-1}g$ is linear on the subcomplex H of L, we may require that $g' = g$ on $|H|$.

(2) If g carries subcomplexes $L_1, \ldots, L_k$ of L into the transversally intersecting non-bounded submanifolds $N_1, \ldots, N_k$ of M, respectively, and if f triangulates a neighborhood of $g(L_i)$ in N_i for each i, then we may require that $g'(L_i) \subset N_i$.

Proof. We may assume that $f^{-1}g$ is a linear isomorphism of H with a subcomplex K, since this may be obtained by a preliminary subdivision. We apply 10.11 to obtain ϵ-approximations $f': K' \to M$ and $g'': L' \to M$, where ϵ is small enough that f' is a triangulation. Then $(f')^{-1}g'': L' \to K$ is linear, and equals $f^{-1}g$ on $|H|$. We let $g' = f(f')^{-1}g''$.

To obtain (2), choose K_i so that $f(K_i)$ lies in N_i and contains a neighborhood in N_i of $g(L_i)$. If ϵ is small enough, $f'(K_i)$ will contain $g''(L_i)$ (see Exercise (b) of 3.11; by 10.11, both lie in N_i). Then $g'(L_i) \subset N_i$.

<u>10.13 Corollary</u>. Let $f: K \to M$ and $g: L \to M$ be smooth triangulations of the manifold M. Given $\delta(x) > 0$, there is a δ-approximation $g': L' \to M$ to g which triangulates M, such that $f^{-1}g': L' \to K$ is linear. If $f^{-1}g$ is linear on the subcomplex H of L, then we may choose g' equal to g on $|H|$.

<u>Proof</u>. If M is non-bounded, this follows at once from 10.12. In the other case, we form the manifold $D(M)$. Assuming the subcomplexes of K and L which triangulate $\text{Bd } M$ to be full subcomplexes, we may double the triangulations f and g, obtaining triangulations $\bar{f}$ and $\bar{g}$ of $D(M)$. We now apply 10.12, taking N to be the submanifold $\text{Bd } M$ of $D(M)$. A restriction of $\bar{g}'$ will be the desired triangulation of M.

<u>10.14 Problem.</u> Let M be a C^r submanifold of N which is closed in N ($M \subset \text{Int } N$). Show there is a C^r triangulation of N which induces a triangulation of M.

<u>Hint</u>: Suppose that both M and N have boundaries. (This is the hardest case.) Extend the imbedding $i: M \to N$ to an imbedding i of a neighborhood V of M in $D(M)$ into $\text{Int } N$. Let f be a triangulation of $D(N)$ which induces a triangulation of $\text{Bd } N$; let g be a triangulation of M. Apply 10.11, letting $N_1 = i(V)$, $N_2 = \text{Bd } M$, and $N_3 = \text{Bd } N$.

REFERENCES

[1] R. H. Bing, Locally tame sets are tame, Ann. of Math., 59 (1954), 145-158.

[2] J. G. Hocking and G. S. Young, Topology, Addison-Wesley, Reading, Mass., 1961.

[3] W. Hurewicz and H. Wallman, Dimension Theory, Princeton Univ. Press, Princeton, N. J., 1948.

[4] J. Milnor, Differential Topology (mimeographed), Princeton Univ., 1958.

[5] J. Milnor, On manifolds homeomorphic to the 7-sphere, Ann. of Math., 64 (1956), 399-405.

[6] J. Milnor, Two complexes which are homeomorphic but combinatorially distinct, Ann. of Math., 74 (1961), 575-590.

[7] E. E. Moise, Affine structures in 3-manifolds, V. The triangulation theorem and Hauptvermutung, Ann. of Math., 56 (1952), 96-114.

[8] D. A. Moran, Raising the differentiability class of a manifold in euclidean space (thesis), Univ. of Ill., 1962.

[9] M. Morse, On elevating manifold differentiability, Jour. Indian Math. Soc., XXIV (1960), 379-400.

[10] J. R. Munkres, Obstructions to the smoothing of piecewise-differentiable homeomorphisms, Ann. of Math., 72 (1960), 521-554.

[11] C. D. Papakyriakopoulos, A new proof of the invariance of the homology groups of a complex, Bull. Soc. Math. Grèce, 22 (1943), 1-154.

[12] N. Steenrod, The topology of fibre bundles, Princeton Univ. Press, Princeton, N. J., 1951.

[13] J. H. C. Whitehead, Manifolds with transverse fields in euclidean space, Ann. of Math., 73 (1961), 154-212.

[14] J. H. C. Whitehead, On C^1-complexes, Ann. of Math., 41 (1940), 809-824.

[15] H. Whitney, Differentiable manifolds, Ann. of Math., 37 (1935), 645-680.

[16] J. Stallings, The piecewise-linear structure of euclidean space, Proc. Cambridge Philos. Soc., 58 (1962), 481-488.

[17] M. Kervaire, A manifold which does not admit any differentiable structure, Comment. Math. Helv., 34 (1960), 257-270.

INDEX OF TERMS